THE
ALIEN INTENT

THE
ALIEN INTENT

A DIRE WARNING

The truth behind the cover-up?

RAYMOND A. ROBINSON

BLANDFORD

A Blandford Book

First published in the UK 1998 by Blandford
Wellington House
125 Strand
London
WC2R 0BB

A Cassell Imprint

Distributed in the United States by Sterling Publishing Co., Inc.
387 Park Avenue South, New York, NY 10016-8810

A British Library Cataloguing in Publication Data block for this book may be obtained from
the British Library

ISBN 0-7137-2732-2

Designed by Anita Ruddell
Printed and bound in Great Britain by MPG Books Ltd, Bodmin, Cornwall

DEDICATION

This book is dedicated to all those people throughout the world who may have suffered an experience of abduction by aliens. Through their insight and painful memories, we may be able to look at our evolution and development in a different way – a way that is devoid of conflict, territorial dispute or a battle of opposing ideologies. Hopefully, by discovering the real reason for the visitors being here, we can collectively rationalize our natural fear of the unknown, and of the change that open contact will inevitably bring.

This book is also dedicated to all those serious researchers of the subject of UFOs, who have often been socially pilloried for their questing activities and beliefs. Without these heretics nothing would happen and we might all be consigned intellectually to the grim darkness of a forgotten history.

CONTENTS

INTRODUCTION

A great many books and articles have been written about UFOs relating the perplexing and often distressing experiences of a growing number of unwilling interactees with the phenomenon. An even greater number of books attempt to explain away or simply list sightings where UFOs or other strange events have been witnessed by countless people. These incidents usually occur quite unexpectedly and without any warning or hint to those involved that something was about to happen which would change their lives forever.

According to some sources, there are currently nine or so alien races visiting Earth. Most of these seem to have no particular interest in us other than that which an extremely foreign tourist might have: that is, to observe, record and take specimens of the flora and fauna. However, there are a few alien types which apparently have a special interest in us as human beings.

Alien intentions are generally manifested in three well-documented ways. The first and most obvious is by giving away their presence in our skies, on our seas and on our land. The second is by abducting people, sometimes from a very early age, subjecting them to various medical procedures, giving them some prophetic message and then releasing them, perhaps to be taken repeatedly in the future. The third is alleged to be sustained institutional contact – co-operation with governments. There is also a less conventional method of contact which takes the form of aliens 'channelling' information to selected individuals, who then are compelled to pass these messages to the general population – the incidence of reported cases peaked in the 1950s and 1960s.

All these events are subjectively real to those who experience them. Sometimes, an 'abduction' may also seem to be a real event to a non-experiencer, for complicated psychological reasons. Unfortunately, because of the tremendous amount of public interest in these strange and increasingly common happenings, people can on occasion delude themselves – there is, after all, a deep-seated human need to be seen, admired and known. It is this conventional aspect of mental health which continues to fascinate many of the psychiatrists faced with such phenomena, who may not discern a real event behind the psychosis of the patient because their prejudice gets in the way. There may, therefore, be innumerable genuine cases of experiencers suffering because of a *real* alien abduction who remain undetected and silent in their confusion and guilt. However, whether an event is *actually real* or merely imaginary is of no immediate relevance to the experiencer, who either way will be very distressed and confused. The fact that the subject *believes* they may have interacted with non-human entities is surely enough for them to seek help and be taken seriously.

In any case, the methods used by the abductors to take people are certainly strange enough to place a question mark against our collective perception of reality. Could our theoretical and quantum physicists be correct when they hypothesize that sub-atomic reality may only exist because of our perception of it? At unobserved times our reality may actually exist as something else. Perhaps there are several realities enmeshed in our own known reality, but eternally hidden from our view.

Many writers have attempted to remove the cloak of secrecy governments have thrown over the whole UFO phenomenon. Timothy Good is a specialist in this subject and he continues to draw back the curtain to inform us of the corrosive effect of secrecy in the name of 'national security'. Great investigative strides have been made in this area, which have culminated in the claim that the US authorities have in their possession actual alien hardware.

Others have tried to explain alien phenomena from an historical viewpoint. Astrophysicist Jacques Vallée is a leader in this field and he has constructed a very convincing argument to show that humans have interacted with non-human entities from time immemorial. Trolls, the jinn, pixies, elves, fairies and other beings have played a part in folk history all over the world, and folklore is brimming with stories of these strange yet normally benevolent creatures. However, all the signs now indicate that humankind collectively may be facing a more threatening presence from outside our Earth. If the pixies and elves are somehow linked to our abductors, it is clear that the mischievous and playful attitude of these delightful fairies has changed for the worse, for reasons not communicated to humankind at large.

Jacques Vallée has postulated that these beings could be intra-dimensional: that is, *they* possess the ability somehow to shift at will between our reality and theirs. This is not as impossible as it may at first seem – at least in theory – particularly when it is now hypothesized with some mathematical conviction that there may be as many as ten different dimensional levels of reality in our Universe, with black holes existing at sub-atomic level and folded dimensions that defy understanding. However, the mechanism which propels *them* between their reality and ours remains outside our understanding for the time being and is likely to do so for the forseeable future.

It is unfortunate that much of the so-called 'proof' of the abductee experience is achieved through the inexact science of regression hypnosis. It is a well-known fact that suggestions can be placed in a subject's mind by a hypnotist who is unskilled in regression techniques: under hypnosis, you are not compelled to tell the truth and there is the constant temptation to say what the hypnotist wants to hear. In such cases, the demography and geography of the abductee should be taken into account: it soon becomes the general perception of any serious student of the subject that most of the modern abductee reports emanate from the US and not, for instance, from so-called Third World countries. In fact, this may not actually be the case, and the perception may have something to do with the fact that the US has a very large population spread over an extremely varied topography, and that culturally more people are willing to risk ridicule by describing their experience and 'coming out'. Nevertheless, the abductee phenomenon remains broadly an American experience.

Many US citizens accept that sentient life probably does exist elsewhere – almost 50 per cent of the population say that they have actually seen a UFO. The rational mind does not, of course, accept this as a bald statement of fact, but it does suggest that a large number of people from any country who share the same conviction may form conclusions based on preconceived views rather than the objective truth. (As far as the US is concerned, the visual wizardry of Hollywood and NASA (National Aeronautics and Space Administration) should also be taken into the inventory.)

Reports of abductions and interaction with non-humans come from all over the world, of course, and it would be wrong to single out the US alone. However, a great deal of material does emanate from there and it is possible to detect a momentum as more information and data becomes available, as though – if the information coming from the abductees is correct – a 'timetable of events' is being played out that may not be under our control.

A very puzzling aspect of most of the reported events concerns the nature of the coded and often meaningless information imparted to the individual during the course of an abduction. Some consider this to be the aliens' way of ensuring that the abduction remains hidden in allegorical screen memories, which exist only as confused agglomerations of unrelated subconscious recollections. The technique can best be described in our terms as 'brainwashing' or 're-programming'. However, despite the thoroughness of the aliens' technique for repressing memory, it is not perfect. The often strange and surreal stories which abductees relate under hypnosis always contain elements that challenge our beliefs – often to the extreme. In cases where there is no evidence to go on other than what the witness says, it must therefore be reasonable to conclude that the witness has perceived what they *say* they have perceived – whether it is *objectively* real or not – until it is substantially proven otherwise.

A recurring theme in so-called 'ufology' is one of disinformation and rumour, circulated by agencies who have vested interests or by individuals who extract some mischievous, spoiling gratification from the chaos they spread. The separation of that which is real from that which is false or exaggerated in any evidence is often very difficult, and is not helped by the fact that many less well-documented incidents are the product of rumour and hearsay – which should automatically exclude such reports from serious investigation.

The abduction scenario is usually inflicted on quite ordinary people who may simply be in the wrong place at the wrong time. The act of describing their experience does not therefore have any direct impact on the society in which they live, but it does allow their story to be told – often at great personal risk of ridicule or derision at local level, and sometimes with catastrophic results on a personal one. In many ways, we are dealing with the stuff of dreams, which is not of our world at all. However, *they* create this reality paradox because *they* intrude and seem to be able to interlock with the deepest parts of the human psyche, which *they* apparently know better than we know ourselves. *They* increasingly interfere in our affairs without our permission: the methods employed by these alien intelligences are truly 'alien' to us.

If there is a timetable or schedule to these events which is under the control of the aliens, it could be hidden in their messages, which commonly take the form of a warning to the human race at large. These are much along the lines of the concerns of the various environmental groups currently operating worldwide, and speak of the human race wrecking the environment through its efficient pillaging of the Earth's resources and escalating levels of pollution. There is also a persistent warning of a possible nuclear holocaust, and there are cases on record of UFOs overflying nuclear missile sites to neutralize warheads, which ties in nicely with the warning message.

It is interesting to note that the aliens are in no way concerned about the current trend towards increased genetic engineering, particularly as it is within the bounds of probability that if new lifeforms are produced by cross gene splicing, nature will seek to create new viruses – against which we will have no immediate defence and which will make the deadly Ebola virus look like a bout of influenza – to adjust the balance. If genetic engineering will allow humankind to produce the food for an increasing world population, it might be reasonable to assume that the aliens have a view on this unrestricted human growth. *They* may also have a view on the subject of the awesome power that any nation controlling gene technology will hold – that is, if *they* have any real interest in our survival at all.

It seems that humankind may currently be under greater threat from gene technology than from nuclear fires. To the best of my knowledge, no warning concerning this aspect of our development has been received from any alien source. It may well be that their timetable ignores it. If it does, its projected timeframe must be relatively short, which makes it crucial to discover what that timetable is. It could also be that there is no timetable at all and we are being tricked into co-operating with the aliens for their selfish ends, for as long as *they* intend to farm us. If aliens are really concerned about the pollution of our planet, because we are using up non-renewable energy sources and employing nuclear power unwisely, it might seem that the message could more pertinently be given to the emerging nations, who in the future will be more prolific users of the remaining fossil fuel resources and cheap nuclear energy than the developed countries, who should be able to afford the emerging new technologies of energy production and conservation.

There are myriad puzzles in the alien phenomenon which we will explore in this book, one of the most important being: why tell ordinary people of impending world doom when they cannot actually do anything about it, other than worry about the consequences? If that is the way the aliens operate in order to change human history, *they* have not learned much from their studies of us humans!

There is a serious side to the matter, and this book attempts to find out more about the aliens' timetable (if it exists) and what their real intentions might be. A word of warning: due to the nature of the subject, much must be conjecture and there will be many paradoxes, because so much of the evidence remains incomplete or apocryphal. A great deal of evidence is nevertheless available, most of it cryptically encoded in the various alien messages and the snatches of memory recall of their contacts. However, most information on the subject (and maybe even the secret of what it is *really* all about) is no doubt locked securely in government vaults. Until recently, most researchers have been content to record or analyse and investigate sighting reports or carry out a programme of exposing government cover-ups. Others de-brief abductees who look for help in unravelling their individual trauma following an abduction, so that they may carry on with their lives as best as they can. The time is right to attempt to discover more about these intruders who invade our air space without permission, and violate our minds and bodies. We need to be in better shape to deal with these unwelcome intrusions – how we, collectively, actually deal with them is a different matter. Understanding what the aliens' *real* intentions are, despite the dense fog of institutional secrecy and fear, must be the first step.

It is sobering to think that increasingly we live in an age which is driven by technology. Since the discovery of silicon-based electronics shortly after World War II, our world has changed forever. Some would say that it has not always been for the better, but either way the impact of solid-state electronics on our world cannot be overstated. Currently, leading-edge research into 'nano engines' has created working machines so small that they are almost at molecular level and could easily be placed inside the human body to perform functions which our biological systems are no longer able to carry out.

Such is the speed at which *Homo sapiens* is progressing scientifically. If we have been 'helped' along the way by our visitors – as some researchers suggest – in order to bring us up to technological speed, it may have been the result of a symbiotic relationship, with visitors and human beings both stakeholders in the outcome. However, current actions on the part of our visitors seems to indicate that the symbiosis no longer exists.

Given all the known data that has been amassed and sifted by government agencies, official bodies and researchers over the last fifty years, it is unwise for anyone to deny that something alien is on Earth. The final proof may come from ordinary people, who have to suffer the indignities of forced physical examinations that amount to rape in which the victim cannot move, scream or protest in any way. It is our collective helplessness in the face of these awful visitations that causes most concern, particularly as our human dignity and everything that we hold sacred is brushed away with cold indifference by *them*, as though we are no more than laboratory animals. Their ability to take us individually to *them* in almost any situation we can imagine should be of the utmost concern, as it implies virtually complete control over our minds as well as our bodies. If this book does no more than to make you think and talk with urgency to others about the phenomenon, it will have done its job.

In writing this book, I have drawn on many sources and have credited other authors or contributors for their valuable input wherever possible. In some instances there is little evidence of ownership to go on, but I have used the data all the same because the message is important. If I have inadvertently left anyone out, I apologize in advance and hope that they will forgive me my omission, and take it in the spirit that 'the sum is always greater than its parts'.

SETTING THE STAGE

Famous Cases

It was in March when the red orbs appeared
A birth of unparalleled redness
When the matchless ruby stars were brought forth
The wife a matchless golden girl...

De stella nova in pede serpenterii: De stella incognita Cygni
Johan Kepler, 1606

THE FIRST ABDUCTION

In UFO lore, it is generally recognized that the earliest modern abduction case concerned Betty and Barney Hill (full accounts of this case are given in *The Interrupted Journey* by John G. Fuller and *Perspectives* by John Spencer – see Bibliography). September 1961 found the Hills returning to their home in New Hampshire from a holiday in Canada. It was late evening and they were near Indian Head when they noticed a bright light in the sky ahead of them, which appeared to keep pace with their car. They stopped the car and, with the aid of binoculars, saw that the light appeared to be a structured, disc-like object with different-coloured lights dancing around its perimeter. Barney eventually left the car to get closer to the object and apparently gained the distinct impression that there were people in the craft looking back at him. Frightened, he rejoined his wife in the car and they drove off at speed. However, once they arrived home they realized that their journey had taken two hours longer than it should have done and they could not account for the missing time. In addition, they discovered twelve or so silver-dollar-sized, shiny magnetic circles on the boot of their car.

Over the next two years the couple apparently suffered recurring nightmares which they attributed to this encounter. They decided to undergo hypnotic regression to try to unearth the cause of their problems and Dr Benjamin Simon regressed the couple (independently and together) once a week for seven months. In assessing the results, it is worth noting that Betty Hill had taken an interest in UFOs between the event and the hypnotic regression, but Barney evidently did not share her enthusiasm.

During the regressions, it was discovered that on the night of their encounter Betty and Barney had been taken from their car and subjected to various medical-type examinations in the landed craft. Betty had had a long needle inserted into her navel area – in 1961 such a procedure was not accepted as a regular medical practice. Later speculation considered the possibility that a sperm sample may have been taken from Barney, fuelled by the fact that he had developed a strange ring of warts on his lower abdomen. However, no admission that a sperm sample had been taken was made at the time. Betty stated that she was shown a star map which revealed where her abductors originated from. One researcher has postulated that the star map places the abductors in the Zeta Reticuli system (see Appendix, page 244), but later investigations have not been so conclusive, especially as Betty Hill had recalled the map from memory. It is also worth remembering that a person under regressive hypnosis will tell the questioner the truth *as they see it* – which may not be the objective truth.

While Barney Hill has since passed on, Betty has continued to experience precognition and paranormal events since the abduction. Dr Simon is on record as stating that he did not believe that the events described actually took place, but a radar report unearthed by investigators Dr J. Allen Hynek and Jacques Vallée confirmed that something was in fact flying in the exact area where the Hills reported their abduction to have taken place. The strange 'beeps' that preceded the abduction itself, and which were also heard after the event, have never been satisfactorily explained; it is suggested that they had something to do with the amnesic method used by the abductors. Even stranger, the circles found on the boot of the Hills' car were never explained either. Investigating these would have been particularly important, because they represented trace features of the event itself, but unfortunately this aspect appears to have been largely overlooked. Overall, the case is a fascinating example of early contact by abduction and is as relevant today as it was when it occurred.

GATHERING SEED

An even more controversial case is the strange tale of Antonio Villas Boas, a twenty-three-year-old Brazilian farmer who worked the land near Fransisco de Sales in the State of Minas Gerais (a full account of the case is given in the *Flying Saucer Review*, Volume 31, No 3). It was October 1957 and Boas was working late (1am) with his tractor in the fields, when he noticed a craft land nearby. He had seen strange lights a few days earlier but had ignored them. When he tried to leave the scene his tractor died on him, and he was eventually taken by force into the landed craft. There he was undressed, sponged with some kind of liquid and a blood sample was taken. Later, a petite and beautiful naked women came into his room and started to caress him. He responded to her attentions and they made love. He recalled that the woman was much like an Earth woman except that she had small pointed features, bright red pubic hair and grunted like an animal while making love – which tended to put him off. Before leaving the room she pointed to her belly, then to Boas and finally to the sky. Presumably, she was trying to tell him that his hybrid child would be born on another world. Villas Boas has, of course, been pilloried for his story, although it is not unique and in 1978–79 two similar cases were revealed – also in Brazil.

In just two cases we have progressed from a medical examination to 'human seed robbing'. The only similarities in the Hills and Boas cases are the fact that the abductors did not feel the need to explain their actions and that it was apparent that there was little the subject(s) could have done to protest at their treatment. This aspect is a continuing thread which runs through all reported contact with aliens. As we shall see, since the Boas case many more similar, if somewhat bizarre, cases have come to light, especially during the 1980s and 1990s.

CLOSE ENCOUNTER?

In 1987, a very strange case emerged from Gulf Breeze, a small peninsula community on the west Florida coastline near the Naval Air Station at Pensacola. The case concerned a local builder named Ed Walters and his family (a full account is given by the Walters themselves in *UFOs: The Gulf Breeze Sightings* – see Bibliography). It is remarkable in that it not only contains accounts of abductions and failed abductions, but more importantly perhaps there are many clear photographs of the craft that Ed Walters saw. These were taken on Polaroid film, which cannot easily be faked or tampered with, and have been subjected to the most critical analysis to verify their authenticity.

The photographs represent some of the most remarkable pictures ever taken of UFOs, in the air and (in one instance) close to the ground. They depict a type of UFO quite different in appearance to the archetypal flying disc typified by the Adamski photographs of the 1960s. There was also the presence of a strange paralyzing blue beam and entities reported in the Walters' house. This case is unique in many respects; it should also be noted that the occupants of the UFO were confident enough to carry out their mission at Gulf Breeze despite the proximity of the Naval Air Station at Pensacola and other flying fields in the area.

Unfortunately, the Walters case suffered a terminal setback with the discovery of models of the UFOs in Ed Walters' possession and the admission of accomplices that the whole thing was a hoax. In addition, it was not helped by the admission that an optical analyst was *paid* to comment on the original Polaroid photographs, rather than their being analysed independently and impartially. Perhaps Walters wanted to 'send up' local believers?

PHOTOGRAPHS AND SAMPLES

A particularly strange case concerns a man named Eduard Albert ('Billy') Meier (a full account of the case is given in *Light Years* by Gary Kinder – see Bibliography). Apparently an uneducated Swiss farmer or casual worker, he claimed to have had contact with 'beamships' from the Pleiades (see Appendix, page 244) since 1976. The photographs taken by Meier near Hinwil in Switzerland were unusually clear and many were in colour; during later analysis, experts found it difficult to fault them. Unfortunately, many of the original negatives taken by Meier were said to have been lost or stolen due to his cavalier practice of giving photographs away and letting his house guests have free rein in his home.

This case is remarkable for many reasons. Apart from the excellent photographs, there is also a recording of the sound of the 'beamship', together with landing traces and a metallurgical sample (which later mysteriously disappeared from an IBM laboratory) to back up Meier's claims. 'Billy' Meier attracted a cult following and visitors to his farm included Shirley MacLaine and many other celebrities.

The Meier case is shot through with the usual false leads and dead ends associated with complicated UFO events. Many claimed that the so-called UFOs were models suspended by wire or string and then photographed. Others (including experts in their field) said that they thought the photographs could not be faked without the resources of a modern Hollywood production team and the assistance of several technical specialists in a studio environment, and it was extremely doubtful that Meier had that kind of facility at his disposal.

The evidence of the missing metal sample is even more tantalizing. The IBM electron scanning microscope showed it to be comprised of tightly bonded, but still discrete, metals and non-metals unknown at the time. An observation that excited the analyst was the discovery of hairline micro-machined grooves joined by furrows in the metal. The major element in the sample was the extremely rare earth element thulium, which had only been produced in minute quantities as a result of the atomic energy work during World War II. The scientist who carried out the analysis concluded that an extensive knowledge of metallurgy would be required even to be aware of such a composition. On further microscopic analysis it was discovered that the sample was at the same time both crystalline and metallic. Unfortunately, the sample then allegedly disappeared in mysterious and, some would say, suspicious circumstances.

Just after the metal sample disappeared and before the photo-journal relating to the case had been published, a Japanese film crew arrived at San José to interview the scientist who had studied the sample. He admitted that he could not explain the material and could not have put it together himself using any technology that he knew of. He had earlier pointed out that while the discovery of the composition of the sample was strange, it did not prove that it was of extra-terrestrial origin. The small metal sample, supposedly from the final stages of production of a 'beamship' hull, remains lost to this day.

'Billy' Meier's case remains fascinating, albeit full of possible hoaxes. It is not helped by the discovery of models of 'beamships' nor by his Greek wife Kaliope's insistence that he was photographing models. It is also not helped by the fact that Meier maintains that he travelled back in time to photograph the celebrated San Francisco earthquake with the help of the Pleiadeans, photographed dinosaurs and met with Jesus, nor by the contention of some that his beautiful 'Semjase' Pleiadean guide and protector/tutor was a girl allegedly picked from a Swedish model directory. Is it possible that Meier invented these wild claims in order to escape the continual bombardment that he and his household had received from followers and other inquisitive people over a long period of time? In order to preserve his photographs and written records, he has now set up an institute called the Semjase Silver Star Centre.

The case of 'Billy' Meier remains largely unresolved to this day, even though an in-depth investigation into it was concluded in 1981.

MILITARY OBSERVATIONS

Another strange event occurred at Rendlesham Forest in Suffolk late in 1980 (a full account of the case is given in *Sky Crash* by Brenda Butler, Dot Street and Jenny Randles – see Bibliography). There were two US Air Force (USAF) bases in the area and a one-page letter/report dated 13 January 1981 on behalf of the Department of Air Force (US) to RAF/CC concerning 'unexplained lights' covered the following:

1 Early in the morning of 27 December 1980 (approximately 0300hrs), two USAF security patrolmen saw unusual lights outside the back gate at RAF Woodbridge. Thinking an aircraft might have crashed or been forced down, they called for permission to go outside the gate to investigate. The on-duty flight chief responded and allowed three patrolmen to proceed *on foot*. The individuals reported seeing a strange glowing object in the forest. The object was described as being metallic in appearance and triangular in shape, approximately 2–3m (6–10ft) across the base and 2m (6ft) high. It illuminated the entire forest with a white light. The object itself had a pulsing red light on top and a bank or banks of blue lights underneath, and was hovering or on legs. As the patrolmen approached the object, it manoeuvred through the trees and disappeared. At this time the animals on a nearby farm went into a frenzy. The object was briefly sighted approximately an hour later near the back gate.

2 The next day, three depressions 2.6cm (1⅛in) deep and 18cm (7in) in diameter were found on the ground where the object had been sighted. The following night the area was checked for radiation. Beta gamma readings of

0.1 milliroentgens were recorded, with peak readings in the three depressions and near the centre of the triangle formed by them. A nearby tree had moderate (0.05–0.07 milliroentgens) readings on the side towards the depressions.

3   Later in the night a red Sun-like light was seen through the trees. It moved about and pulsed. At one point it appeared to throw off glowing particles and then broke into five separate white objects and disappeared. Immediately thereafter, three star-like objects were noticed in the sky, two to the north and one to the south, all of which were about 10 degrees off the horizon. The objects moved rapidly in sharp angular movements and displayed red, green and blue lights. Those to the north appeared through an 8–12 power lens to be elliptical. They then turned into full circles. The objects to the north remained in the sky for an hour or more. The object to the south was visible for two or three hours and beamed down a stream of light from time to time. Numerous individuals, including the undersigned Lt Col Charles Halt, witnessed the activities described in paragraphs 2 and 3.

The usual red herrings were present in abundance in this case, from tape recordings of the actual events to eyewitness accounts of a meeting with small alien creatures by the base commander. Explanations for the happenings in the woods ranged from the practising of satanic rites to drug abuse by air force personnel and the consequent coverups that would ensue. The case has been the subject of television documentaries and continues to generate many books. It is strange but probably significant that the case only really came to light through access to US Air Force files via the Freedom of Information Act (FOIA). The British Ministry of Defence (MOD) were already busy denying that anything had happened at all!

ROSWELL

Perhaps the most famous, most documented and possibly most notorious case of all concerns the so-called 'Roswell' incident. The events which took place on or around 3 July 1947 in the New Mexico desert have continued to attract much media attention over the decades, with a general-release film being the latest in a long line of investigative works undertaken in an attempt to get the US government to admit that a massive cover-up was mounted to mask a real event. (The case is depicted in the films *Hanger 18* and *Roswell – the Film*, and a full account is given in *UFO Crash at Roswell* by Kevin D. Randle and Donald R. Schmitt – see Bibliography).

In outline, it is claimed that something otherworldly crashed at Roswell *and* alien occupants were recovered. A rancher was supposedly forced, in the name of patriotism, to lie about the unearthly debris he had found on his ranch, and a Major Jesse A. Marcel of the 509th Bomb Group of the Eighth Air Force (the Atomic Bomb Group responsible for dropping the A-bombs on Japan) at Roswell Army Airfield was ostracized and ordered to state publicly that he had misinterpreted the wreckage, which was in fact that of a weather balloon – notwithstanding that it was unlike any weather balloon he had ever seen. Wreckage was scattered over a very wide area; and as the bulk of it was not in one vicinity, search aircraft were sent up to see if anything could be seen from the air. The search crews allegedly found the remains of a circular aeroform – a disc – some kilometres from the original pieces of debris.

More sensationally, it was said that there were 'humanoid' casualties. Some accounts state that there were four dead humanoids, others that there were three dead and one survivor, who was found tending a dying colleague. The survivor was said to have panicked totally when armed military personnel appeared and had to be restrained.

A strange inconsistency concerns a group of archaeologists who were allegedly first on the scene where the disc had crash landed – despite the fact that scientists interested in the history of New Mexico had said there was nothing of interest in the area to draw the archaeologists there in the first place.

The passage of time has, of course, dimmed memories, but although the enormity of Roswell was made clear to witnesses by military threats of national security breaches and the severe penalties they would incur if they opened their mouths, a number of deathbed confessions have confirmed the main thrust of the incident.

To this day, some still hail Roswell as the most evidential case for a crash/retrieval of an alien flying disc. It was said that the pieces of the crashed disc were taken by B29 to Wright Patterson Air Force Base, presumably for analysis and interpretation. The alien remains were apparently autopsied, photographed and logged. Legend has it that the remains were housed at the base in what was euphemistically called the Blue Room. Some sources maintain that there were actually *two* crashed discs, one in the Roswell area and the other at Aztec, following a collision in mid-air. However, the US government still insists that nothing unusual happened at Roswell in July 1947, and with the passage of time this complicated case has become almost impossible to unravel.

More recent research suggests a more mundane reason for a cover-up: the Roswell disc may have been invented to disguise the top secret Project Mogul. This was essentially a high-altitude listening device designed to detect high-yield explosions – atomic blasts – anywhere in the world and was particularly relevant at the time, when nations were obsessed with the Cold War. Couple this with the fact that the poor farm tenant, Mac Brazel, had allegedly heard of a bounty being given for the capture of a flying disc, and you may have a recipe for deception. Nevertheless, the Roswell incident still receives an enormous amount of publicity, with sensational new claims being made from time to time.

PUERTO RICO – HOT SPOT

The cases covered so far have ranged from abductions which seemed to have been used (at least initially) to find out what makes a human being 'tick', through contacts where information and enlightenment ensue, to crash retrievals where we, humankind, can get our hands on alien technology, even if we find it difficult to understand or duplicate. The crash retrieval theme is very important, because it proves to us that *they* are mortal too – which is something we can identify with, albeit negatively. It also proves that their craft are 'nuts and bolts' machines.

An exciting, but more sinister, twist concerns the island of Puerto Rico, which is a US territorial possession in the Caribbean. Leading investigator Jorges Martin has catalogued many significant UFO-related events on the island, focused specifically in the south-west region and most particularly the Laguna Cartegena area (a full account of this is given in *Alien Update* by Timothy Good – see Bibliography). Considering that there is a US military radar blimp facility on the island and that it is also the home of SETI (Search for Extra-terrestrial Intelligence – the huge static radio dish at Arecibo on the Atlantic ocean side of the island), the number and type of UFO events that continue to occur here is quite remarkable.

There have been cases of underground 'explosions', resulting in collateral damage through the resultant earth tremors. Huge flying discs have been seen to hover over the 'earthquake' epicentres. US military jets have been seen chasing (or perhaps escorting?) discs around the Cabo Rojo area. Witnesses have seen the jets being 'absorbed' by huge flying objects. Inhabitants have been frightened by seeing discs flying out of lakes – particularly Laguna Cartegena. Humanoid beings have even been seen walking down a road.

Strangely, all these happenings seem to be known to the US military, who are regularly seen in areas associated with unusual events. Witnesses have claimed that in May 1987 they saw strange lights descending into Laguna Cartegena, followed by a huge disc hovering low over the water as though searching for something. The next day an enormous tremor and explosion were felt which rocked the whole region. One of the witnesses saw cobalt blue-coloured smoke rising from cracks in the soil and was so frightened that he called in specialists from the University of Puerto Rico, who supposedly were going to investigate the matter. Strangely, they did not even bother to take samples of the smoke, nor of the powder that remained on some of the contaminated plants. The blue smoke apparently also issued from Laguna Cartegena itself and the lake was subsequently cordoned off by unidentified personnel, some of whom were clothed in camouflage fatigues. Others were dressed in plain clothes, and some wore white anti-contamination suits with dark grey gloves and boots and were engaged in collecting water, mud, soil and plants, presumably for examination. It is also believed that radiation counts were taken of the area.

The following day an unmarked dark green helicopter flew over the area and a large metallic ball was lowered into the lake as though something was being searched for. There were also a number of vans and four-track vehicles with small parabolic, revolving antennas on their roofs. When a local woman and her sister went to the area, they were ordered to leave and were told that the men at the lake were trying to establish what had happened – if an earthquake had caused the damage and the blue smoke, why would they have been told that the cause of the explosion and resultant tremor was unknown? The same woman admitted to Jorges Martin that she had seen discs entering or exiting the lake since 1956! She also stated that, out of curiosity and bravado, her brother (now deceased) had sought to find out whether or not, as rumour had it, the occupants of these discs came from outer space. He visited the lagoon at night and had an encounter with two white men who were 2–2.2m (6–7ft) tall. Both had long blonde hair and were dressed in one-piece, tight-fitting silvery suits. The witness said later that they were very beautiful and delicate, almost feminine. At his insistence – he was very nervous – after smiling at him they retreated and walked back into the lake. Unfortunately, the witness was not believed by his family and he did not mention anything else regarding the aliens. However, his sister was sure that he had several more encounters without telling anyone of them, for fear of ridicule.

The US 'radar' blimp is also the subject of attention from UFOs. The official purpose of the blimp was given as the control of drug smuggling by land and sea in the immediate area. However, on numerous occasions since its installation local people have seen UFOs hovering next to the blimp and emitting similar strobing flashes to its own, as though some sort of communication is in progress. Sometimes US military jets chase the UFOs away, but every time a UFO appears next to the blimp the latter malfunctions and has to be repaired. According to some sources, all the computer systems go down and then have to be re-initialized.

As a result of all this, many believe that the blimp has nothing to do with drug smuggling but instead is used to monitor the mischievous visitors. Some of the residents also believe that the aliens have a base on the island and the US military presence is there to ensure that they stay where they have been put. The fact that US federal agents are in evidence indicates that the US government know what is going on in Puerto Rico.

Flying with the aliens

A quite remarkable experience befell Carlos Manuel Mercado one night in June 1988. Unable to sleep because of the heat, he got up from his bed, went to his living room and lay down on the sofa to try to get some sleep. Suddenly, he was aware of a flash of bright light outside and heard a buzzing sound. Not long after this he heard knocks on his venetian blinds. On getting up to investigate, he came face to face with three small non-human entities. He heard a voice in his head (telepathy?) which said, 'Do not be scared, everything is fine. Nothing is going to happen to you. Do not be scared. We want to show you something.' Feeling passive and unable to resist, Manuel opened the door and accompanied the three small humanoids down the road to a waiting 'flying saucer'. He noticed that the beings were about 1.2m (4ft) in height with large pear-shaped heads, pale skin and big black slanted eyes with no pupils. They had no ears, a small slit for a mouth and just two holes for a nose. They did not open their mouths when they 'talked' to him. Rather curiously, Manuel remembered that they had small bumps in the skin of their faces, which he suggested was akin to acne. Their arms appeared to be longer than a human's and they were all dressed in what appeared to be close-fitting, sandy-coloured one-piece suits. Only their hands and heads were outside their clothing.

The 'flying saucer' stood on three legs. It was circular and had a dome on top with windows. The base had many coloured lights around the rim. An opening on the underside revealed a hatch from which a long ladder extended to the ground. The humanoids asked Manuel to climb aboard the craft. He saw more small humanoids inside, but was introduced to a taller being closer to Manuel's own height, 1.7m (5ft 9in), who was dressed in a white robe. He felt more at home with this creature, as he looked more 'human' than the others; Manuel was given to believe that this taller alien was the 'Captain-medic'.

The flight deck was full of machinery with panels and lights. The tall entity told Manuel that they meant him no harm, they just wanted to show and tell him things that he could relate to others. The being said something to the smaller humanoids and Manuel heard the landing legs come up with a clamping sound. The hatch door closed and the craft started to move, but somehow the aliens prevented him from becoming fearful. The craft shot up into the sky and Manuel thought he would be transported far away. However, it veered left and descended towards Sierra Bermeja. Manuel remembered being afraid that the craft would crash, but a hole appeared in a depression to the side of El Cayul mountain and the craft went down into a tunnel, emerging into what appeared to be a long cavern. Manuel saw barrack-type structures and hundreds of small aliens working as though on production lines, assembling electronic and mechanical parts for machinery. There were many strange craft in the cavern which were saucer-shaped, triangular or hexagonal.

The tall entity told Manuel that they have a base here for the maintenance of their craft's systems. It said that they had been there for a very long time and do not intend to leave. They mean no harm to Earth people and do not want to conquer us. Instead, they want to reach out to us and establish a direct relationship which would be beneficial to both parties. The fact that Earth people were meant no harm was emphasized once again.

Manuel asked why he had been selected for this mission, for he was a simple man and no one would believe him. The entity said that it did not matter, as they were contacting several others and he would be just as likely to be listened to as any of the others. They then took him home and promised that they would return someday. Manuel kept all of this to himself until he heard about the US jets being absorbed and then decided to talk.

Interestingly, another man residing in the same area of Puerto Rico was taken to the same base by the same type of alien beings. The man is a high-ranking military officer and his story corroborates Manuel's.

It is clear that something important is happening in Puerto Rico. Over the years there have been numerous witnessed incidents of disc-shaped craft flying at will around the island. There have been cattle mutilations, abductions and apparent US military involvement. The Director of the Civil Defense Agency of Lajas, Freddie Cruz, witnessed a jet fighter (possibly an F14 Tomcat) chasing a disc-shaped UFO on the afternoon of 28 April 1992. The disc looked like two flattened dishes with a dome on the top; it was 'silvery', polished and slightly bigger than the jet. It continually stopped in the air and then, just as the jet seemed to catch up, it would accelerate and then stop further away. The jet was about to close in again when the disc split in half; one half went south-west, the other east. The jet remained in the vicinity, circling the area as though unsure of what to do next, until finally departing to the east.

Freddie Cruz is convinced that something very strange is going on on his island. He claims that white NASA trucks enter the radar installation at about 02h00 escorted by military jeeps. He asks, somewhat rhetorically, 'What have NASA got to do with the anti-drug war?'

The remarkable events in Puerto Rico continue to be the subject of investigation by Jorges Martin and it is likely that more details will eventually emerge. In fact, it is very difficult to see how the lid can be kept on such amazing events for much longer.

DISAPPEARANCE I

To recap: the cases reviewed so far have described abductions for medical purposes, with the possibility in mind of sexual breeding with an alien race. Contact accounts are given so that the contactee's experiences may be passed on for humankind's general edification. The discs' occupants seem to delight in their displays of clear aerial superiority, with feats of technical wizardry approaching the status of miracles. The aliens appear to operate with or without any co-operation from us, although it is probably beneficial to them if interference is kept to the minimum while they carry out their unknown agenda. The cases covered are just the tip of a worldwide metaphorical iceberg. More disturbing and bizarre events do occur and include cattle mutilations, and in some cases human deaths. However, returning to the abduction scenario, virtually all the subjects are merely 'borrowed' for a period of time, the longest being a few hours. But there are instances where the abductee is never seen again, and the Australian case of Frederick Valentich may be one.

Frederick Valentich was a twenty-year-old flying instructor who disappeared with his Cessna 182 light aircraft shortly after reporting a UFO over the Bass Strait near Cape Otway. He was on a flight from Moorabin, Victoria, to King Island, Tasmania, on 21 October 1978. Almost fifty minutes after leaving Moorabin Airport, Melbourne at 1819hrs, Valentich radioed to the Melbourne Flight Service Unit Controller that he had sighted an unidentified aircraft. The following is a transcript of the official flight recording tape provided to Timothy Good by researcher Bill Chalker and is reproduced here with the kind permission of Timothy Good from his excellent reference work *Above Top Secret* (see Bibliography). The recording of the communications between the aircraft (VH-DSJ) and the Melbourne Flight Service Unit (FSU) commenced at 1906hrs. Words in parentheses were indistinct and are open to other interpretations.

Time	From	Text
1906:14	VH-DSJ	MELBOURNE this is DELTA SIERRA JULIET is there any known traffic below five thousand?
:23	FSU	DELTA SIERRA JULIET no known traffic
:26	VH-DSJ	DELTA SIERRA JULIET I am seems (to) be a large aircraft below five thousand
:46	FSU	DELTA SIERRA JULIET what type of aircraft is it?

:50	VH-DSJ	DELTA SIERRA JULIET I cannot affirm it is four bright it seems to me like landing lights
1907:04	FSU	DELTA SIERRA JULIET
:32	VH-DSJ	MELBOURNE this (is) DELTA SIERRA JULIET the aircraft has just passed over me at least a thousand feet above
:43	FSU	DELTA SIERRA JULIET and it is a large aircraft confirm
:47	VH-DSJ	er... unknown due to the speed it's travelling is there any air force aircraft in the vicinity?
:57	FSU	DELTA SIERRA JULIET no known aircraft in the vicinity
1908:18	VH-DSJ	MELBOURNE its approaching now from due east towards me
:28	FSU	DELTA SIERRA JULIET...
:42		//open microphone for two seconds//
:49	VH-DSJ	DELTA SIERRA JULIET it seems to me that he is playing some sort of game he's flying over me two to three times at a time at speeds I could not identify
1909:02	FSU	DELTA SIERRA JULIET roger what is your actual level ?
:06	VH-DSJ	my level is four and a half thousand four five zero zero
:11	FSU	DELTA SIERRA JULIET and confirm that you cannot identify the aircraft
:14	VH-DSJ	affirmative

:18	FSU	DELTA SIERRA JULIET roger standby
:28	VH-DSJ	MELBOURNE DELTA SIERRA JULIET its not an aircraft it is... //open microphone for two seconds//
:46	FSU	DELTA SIERRA JULIET can you describe the er... aircraft?
:52	VH-DSJ	DELTA SIERRA JULIET as it's flying past it's a long shape... //open microphone for three seconds// (cannot) identify more than (that it has such speed)... //open microphone for three seconds// before me right now MELBOURNE
1910:07	FSU	DELTA SIERRA JULIET roger and how large would the er... object be?
:20	VH-DSJ	DELTA SIERRA JULIET MELBOURNE it seems like its stationary what I'm doing right now is orbiting and the thing is just orbiting on top of me also it's got a green light and sort of metallic (like) it's all shiny (on) the outside
:43	FSU	DELTA SIERRA JULIET
:48	VH-DSJ	//open microphone for five seconds// it's just vanished
:57	FSU	DELTA SIERRA JULIET
1911:03	VH-DSJ	MELBOURNE would you know what kind of

		aircraft I've got is it (a type) military aircraft?
:08	FSU	DELTA SIERRA JULIET confirm the er... aircraft just vanished
:14	VH-DSJ	say again
:17	FSU	DELTA SIERRA JULIET is the aircraft still with you?
:23	VH-DSJ	DELTA SIERRA JULIET (its ah... nor)
		//open microphone for two seconds//
		approaching from the south-west
:37	FSU	DELTA SIERRA JULIET
:52	VH-DSJ	DELTA SIERRA JULIET the engine is rough idling I've got it set at twenty-three twenty-four and the thing is (coughing)
1912:04	FSU	DELTA SIERRA JULIET what are your intentions?
:09	VH-DSJ	my intentions are ah... to go to King Island ah... MELBOURNE that strange aircraft is hovering on top of me again...
		//open microphone for two seconds//
		it is hovering and it's not an aircraft
:22	FSU	DELTA SIERRA JULIET
:28	VH-DSJ	DELTA SIERRA JULIET MELBOURNE
		//open microphone for seventeen seconds//
:49	FSU	DELTA SIERRA JULIET MELBOURNE

No further transmissions from the aircraft were recorded. Despite the weather being perfect for flying, no trace of Valentich or his plane were found, even after search-and-rescue procedures were declared at 1912hrs when he did not arrive at King Island. The search for traces of the Cessna continued until 25 October 1978 but nothing was ever found. (Stories circulated at the time that a wreck had been discovered on the sea bed, but these were generally considered hoaxes.)

Interestingly, the complete tape was in the possession of Dr Richard Haines, a NASA research scientist whose preliminary investigations indicated that there was a strange seventeen-second burst of metallic noise following Valentich's last transmission. The noise contained thirty-six separate bursts, with fairly constant start and stop pulses bounding each one. There were no discernible patterns in time or frequency. The noise was similar to the rapid keying of a microphone, but control simulations did not yield the same sound as the original noise on the tape.

The official verdict of the event concluded that Valentich was probably dead and that the cause of the accident was unknown. Many sightings of strange aerial objects were witnessed on the same day that Valentich disappeared from Australian skies. Over twenty years after the event, witnesses are now coming forward stating that they saw Valentich's Cessna with a green light immediately above it at the time of his disappearance. However, the case remains an unsolved mystery and a tragedy for his family.

The Valentich case is not typical of what could be considered a 'normal' abduction event (if there is such a thing), and is one of the very few cases where a person is presumably taken and is never seen again. It is a well-known fact the area around the Bass Strait in Australia (an approximate 320km (200-mile) stretch of water between Tasmania and Australia) is a UFO 'hot spot' and it may well be that the unfortunate Frederick Valentich fell prey to a regular patrol craft.

DISAPPEARANCE II

The tragic case of Captain Thomas Mantell, a war veteran, on 7 January 1948 is worth recalling, as accounts of the incident appear to differ or omit various details. The commonly accepted view today is that he died in his P51 Mustang as a result of chasing either the planet Venus or a weather balloon, or that he suffered asphyxiation due to the fact that his climb exceeded 6,750m (22,000ft) and he had no on-board oxygen supply.

However, a fuller explanation appears in Donald Keyhoe's book *Flying Saucers from Outer Space* (see Bibliography). Following Ken Arnold's historic sighting of 'flying saucers' in 1947, there were waves of sightings all over America. Experienced air force personnel saw silver discs circling Muroc Air Force Base. Commercial airline crews saw discs in Idaho air space.

While AFB Commanders kept their bases on alert for fear of a Russian secret weapons invasion, Washington officials played down the sightings as hallucinations or optical aberrations. It is alleged that a small group of intelligence officers urged the USAF to set up an investigation, presumably to find out what was going on once and for all. Unfortunately, Captain Mantell's death seems to have interrupted their plans.

On the afternoon of 7 January 1948 a huge circular, glowing object was seen by hundreds of witnesses at Madisonville, Kentucky. The sighting was later confirmed as having been seen by people right across the State. Fort Knox was warned by the Kentucky State Police, who estimated the object's size to be a minimum of 77m (250ft) in diameter. Half an hour later the object appeared over Godman AFB near Fort Knox. It hovered over the airfield, glowing alternately red and white.

Captain Thomas Mantell and three other pilots were on a training flight when they flew over the airfield. Because they were already in the air and in close proximity, they were asked to investigate the object. A few minutes later, having climbed through broken cloud formations, Mantell reported that he thought it was metallic and that it was of tremendous size. He commented that the object had now started to climb away from him, but he would follow it up to 6,150m (20,000ft). He confirmed that he would abandon the chase if he was no closer to the object at that altitude. Minutes passed and the control tower hailed Mantell again – no reply.

Later the same day Captain Mantell's body was found near his wrecked plane, around 145km (90 miles) from the airfield. A witness said that the plane had seemed to explode in mid-air. It was said that the wreckage showed no signs of fire, but it was established that it had disintegrated before it hit the ground.

The following day the press reported that it had been discovered that Mantell's body had been struck by a type of ray, or that he had been whisked away by spacemen. The air force refused to release any details or pictures of the crash at the time, although out of respect for Captain Mantell's relatives rather than for any clandestine reason.

INVESTIGATIONS

Soon after Captain Mantell's death, the air force initiated Project Sign (press name Project Saucer), the first serious investigative committee which would try solve the riddle of the saucers. The committee was comprised of intelligence officers, rocket experts, aeronautic engineers, an astrophysicist and several other scientists. At its inception the project was top secret.

Donald Keyhoe later published a magazine article speculating that the discs involved were piloted by extra-terrestrials (ETs) who were visiting Earth. The USAF was clearly sensitive to such claims and took the unprecedented step of inviting a staff writer from a news service to an interview with a Major Jere Boggs (a project intelligence officer who served as Liaison Officer between the Pentagon and Wright Field AFB, an Air Technical Intelligence Center – ATIC). The interview followed a flat denial that the saucers even existed!

During the interview, Boggs was asked about Mantell's death and answered that Mantell had been fooled by the planet Venus. Later, Keyhoe managed to interview Boggs himself. He asked him again about Mantell's death and Boggs reaffirmed that Mantell had been chasing Venus when he was killed. Keyhoe remarked that Venus was practically invisible on the day in question and that Boggs' statement was in direct contradiction to the April project report, which after some fifteen months had concluded that the object Mantell had chased was not Venus. Boggs replied that after a reappraisal of their original conclusions the project had decided that the object had, in fact, been Venus.

It appears that Keyhoe had discovered that the Pentagon had been cornered and had had to make quick denials in order to maintain the security status quo. Unfortunately for Boggs, Wright Field AFB confirmed that they had not said that the object was Venus. Somehow, Project Sign had released case studies and sightings which by that time had been de-classified and were not secret any more. Keyhoe managed to get hold of the summaries and was surprised to see that the summary of the Mantell incident concluded that:

'Under exceptionally good atmospheric conditions, and with the eye shielded from the direct rays of the Sun, Venus might be seen as an exceedingly tiny bright point of light. However, the chances of looking at just the right spot are very few.

It has been unofficially reported that the object was a naval cosmic ray research balloon. If this can be established, it is to be preferred as an explanation *(later to be proven false)*. However, if one accepts the assumption that reports from various other localities refer to the same object, any such device must have been a good many miles high in order to have been seen clearly, almost simultaneously, from places 280km (175 miles) apart... no man-made object could have been large enough and far away enough for the approximate simultaneous sightings.

It is most unlikely, however, that so many separated persons should at that time have chanced on Venus in the daylight sky... The sighting might have included two or more balloons (or aircraft) or they might have included Venus (in the fatal chase) and balloons... Such an hypothesis, however, does still necessitate the inclusion of at least two other subjects than Venus, and it is certainly coincidental that so many people would have chosen this one day to be confused (to the extent of reporting the matter) by normal airborne objects.'

Based upon what he had learned from the Mantell case and others, Donald Keyhoe was now able to conclude that:

1 The USAF was puzzled and some officials were worried by the volume of reports during 1947.

2 The USAF suspected the truth about the discs after Mantell's death – if not before.

3 Project Sign was created to investigate, and at the same time conceal, the truth.

4 Project Sign policy had been set by Defense Secretary James V. Forrestal (see page 146) in 1949, only to be reversed by the Pentagon. Someone had decided to release the facts gradually in order to prepare the American people.

5 Keyhoe's first saucer article had been considered part of this gradual education exercise but public reaction scared USAF officials, causing the denials that the saucers were real.

6 Major Boggs had been told to push the Venus solution to Mantell's death to avoid closer examination of the case. It was denied by Project Sign, as the USAF knew that most people had either forgotten or had never known about it. Keyhoe speculated that Boggs had to stick to his story because he did not know that the case summaries would be released into the public domain.

7 Case summaries were shown to a small, select group of Washington newsmen to continue the ET theme.

Prophetically, Keyhoe continued:

'I believe that the Air Force is still investigating the saucer sightings, either through the Air Materiel Command or some other headquarters. It is possible that some Air Force officials still fear a panic when the truth is officially revealed. In that case, we may continue for a long time to see the routine denials alternating with new suggestions of interplanetary travel.'

Nothing much seems to have changed since Donald Keyhoe wrote those words. Predictably, Project Blue Book (1952–69), last in the line of USAF projects emanating from the generic Project Sign and including Projects Grudge, Twinkle, possibly Blue Paper etc, concluded at some length that Mantell had been chasing the planet Venus.
Keyhoe went on to try to disentangle the saucer problem by speculating as to why the saucers were here (my comments in italics):

1 The unknown ET race (perhaps more than one) might fear an invasion by *Homo sapiens* once space travel has been conquered. Success with nuclear detonations and high-altitude rockets may have escalated the aliens' fear. This hypothesis was one of a number of conclusions reached by Project Sign. *Strangely, the modern view of Stanton Freidman seems to be in some accord with that theory, inasmuch as he has postulated that the aliens are here in order to stop us escaping from Earth.*

2   The ETs fear more powerful nuclear explosions if they originate in our solar system.
Is it possible that a number of simultaneous H-bomb explosions could speed up the Earth's orbital velocity or even change its orbit? Dr Paul Elliott, a physicist who worked on the first H-bomb, was an adherent of that theory. Others thought that the Earth might suffer a mortal blow, thereby having a possibly fatal knock-on effect on the other planets in our system.

Donald Keyhoe then contributes some more of his own suggestions, as well as some that he had gleaned from others:

3   The ETs use nuclear energy and are running out of raw materials. The nuclear detonations on Earth reveal that we have discovered uranium and the ETs have therefore zeroed-in on our supplies. Keyhoe speculates that more saucers are seen over the US because it is the most advanced nation in this field.

4   The saucers may be part of an ET programme to conquer inhabited planets.

5   ETs may have some unknown plan for Earth besides conquest.

6   The Earth may not be considered a menace to ETs and might not hold any material interest which would lead to an invasion. Surveillance may therefore be:

(a) to survey the Earth with the intention of contact once the ETs have convinced us of their peaceful intentions and are sure we will not attack them

or

(b) to catalogue the Earth as just another inhabited planet with no immediate plan for contact. The ETs might be carrying out periodic checks to make sure that humankind poses no threat or that we seem advanced enough for acceptance by *them*.

It is hard to imagine what the atmosphere might have been like in those days of international high tension. Following World War II, the US remained on high alert. Stories of Russian experiments from captured secret Nazi flying-disc prototypes only served to heighten the tension. The sudden appearance of strange discs all over the US, particularly around Washington, and the inability of the military might of a nation that had just won a world war to control its own air space probably laid the foundations for the official paranoia that exists today. As far as the programme of official release is concerned – it never really happened, unless it was given to Hollywood, which was more suited to dealing with fantasy than suggestions of fact.

As far as the unfortunate Captain Mantell is concerned, it seems that the truth behind his death will have to wait a few more years until all the skeletons are out of the cupboard before it is revealed. Although his was not a case of abduction, there is no doubt that some kind of accident did occur which resulted in the death of a veteran pilot in suspicious circumstances, which the USAF (or more probably the Pentagon) made worse by denial and hastily cooked-up excuses which insult the public's intelligence. The incident is therefore very important historically, particularly as many would consider that the jury is still out.

ALIEN CHILDREN?

F inally, the strange and harrowing case of Kathie Davis (real name Debbie Tomey) should be mentioned. In collaboration with UFO researcher Ted Bloecher and psychologist Dr Aphrodite Clamar, the pioneering American investigator Budd Hopkins had written a book called *Missing Time: A Documented Study of UFO Abductions*, detailing the cases of seven reported abductions. As a result of reading the book, Kathie Davis contacted Hopkins, believing that she was suffering from an unremembered trauma concerning an abduction.

Kathie Davis's case was investigated using regression hypnosis, voice-stress analysis and psychological tests over a period of some two-and-a-half years. As the case unfolded, it transpired that Kathie had experienced several abductions and had been instrumental in providing ova (taken from her by the aliens) for a breeding programme – presumably an experimental but limited one – over a long period of time. Kathie met and fell in love with her husband-to-be in 1977. They planned to marry in 1978, but Kathie realized that she had become pregnant, which the usual tests confirmed.

Accordingly, the couple put back the date of their wedding by a few months. One day, Kathie awoke with what appeared to be a normal menstrual flow, and a visit to her doctor soon confirmed that she was no longer pregnant. There had been no miscarriage, nor any of the normal physical traces one would expect from a natural termination. Kathie was just not pregnant any more. She went on to have two children, but the loss of that first child would always be a tragic memory for her.

Under hypnosis, Kathie remembered being abducted and that she had been used in a breeding programme. She remembered being given a small, frail infant to hold while in an alien craft. The infant seemed to be a mix of alien and human, and she was given to believe that it was part of her. She was shown a young girl, perhaps around four or five years old. The child had unusually pale skin and wispy, unkempt hair which barely covered her large skull. Kathie was given to believe that this child also had something to do with her. She remembered the strange sensation that the aliens were somehow learning from these sad and pathetic meetings with the 'children'. She sensed that *they* hoped to learn how a mother feels love and tenderness. This was something *they* did not possess, which was why *they* needed Kathie to show *them* how tenderness could be demonstrated. *They* appeared not to know how to suckle a baby (even if *they* were capable) until Kathie demonstrated the technique.

The aliens' inability to understand the relationship between mother and child tells us much about their lack of understanding of the condition of being human. This lack is remarkable, especially when we consider how advanced *they* are in their apparent use of technology, which is miraculous when compared to our own brute force methods. Perhaps it is the unpredictable nature of the complex interplay between the human self and the aliens' stimuli which *they* do not currently understand fully enough to make a permanent contact viable with no risk to their survival in our company, or indeed to ours in theirs.

Whether the subjective weaves of believers and activists like Donald Keyhoe move us forward in understanding the UFO phenomena remains debatable; whether the natural instincts of a mother or the tragic death of an experienced pilot bring us nearer to the truth of the abduction experience is also very much open to question. However, there is still much to consider. It is time now to leave the subject of abduction from the point of view of the abductees and attempt to examine the alleged perpetrators of these acts.

THE ALIENS

Beholding from afar I saw a great cloud loomimg black over all the earth, that had absorbed the earth, which covered my soul.

St Thomas Aquinas

It has been said that there are currently nine species of alien visiting our planet. The main types that have been recorded from witness accounts are:

1   Nordic types with long, flowing blonde hair, normal human height, handsome features and normal human physical characteristics.

2   Giants 2.2–2.7m (7–9ft) tall, in correspondingly large ships and hardly ever seen outside their vessels.

3   Reptiles with large, gruesome reptilian heads and scaly skin, usually seen in conjunction with other alien types when performing some work function involved with an abduction.

4   Monsters, usually with what appears to be breathing apparatus attached to their heads. Not all 'monsters' appear to have what could be called a 'head', some have brightly glowing red eyes. Non-oxygen breathers.

5   Goblin-like entities with huge ears and very long arms.

6   Robots or remotely controlled objects, which can be any configuration from spiked ball-type devices to humanoid patterns.

7   'Greys', 1–1.2m (3–4ft) tall, with grey to pasty-white skin, long thin arms, three fingers (with claw-type endings), thin bodies and large pear-shaped heads with almond-shaped, wrap-around black eyes, hardly any nose, a thin slit for a mouth and vestigial ears. Communication between *greys* and humans is non-verbal.

8   Tall, very thin, ephemeral-type beings with very long arms, who seem to be 'in charge' of smaller 'greys'. Often seen in a 'floating' type of motion.

9   Bigfoot-type beings.

NORDICS

Perhaps the earliest mention of the 'Nordic' type was by George Adamski, as a result of his alleged meeting with one in the Californian Desert (a full account of this encounter is given in *Flying Saucers Have Landed*, by Desmond Leslie and George Adamski – see Bibliography). A similar type of alien was mentioned by 'Billy' Meier (see page 18), although in the former case the alien said he came from Venus, Earth's sister planet, and in the latter from the Pleiades star system. Unfortunately, it is now known that Venus cannot sustain intelligent life as we know it, and it is also very unlikely that the Pleiades, as a comparatively young star system, would have evolved a similar planetary system to our own. Other 'Nordic' types have been seen comparatively recently in Puerto Rico.

The 'Nordics' are frequently mentioned by witnesses. It is worth noting that these witnesses almost always state that they sense nothing but love and understanding emanating from this type of alien. It appears that all *they* want to do is to give us greater knowledge about life and the Universe and our collective place in it, so that we may live in greater harmony with our stellar brothers and sisters. There is no mention of the forced taking of eggs or sperm by this type of alien.

GIANTS, REPTILES AND MONSTERS

Giant aliens have been reported mainly in what used to be Soviet Russia and, more peculiarly perhaps, the Canary Islands. In the former case *they* have been seen outside their vessels and have only succeeded in scaring the local population witless, while in the latter, there have been reports of *them* being seen suspended in what appear to be gigantic iridescent soap bubbles (in all such accounts it must be remembered that witnesses struggle to rationalize what they have seen and will resort to descriptions of familiar objects to describe an unnatural and frightening experience).

The 'reptile' form seems only to be seen occasionally by some abductees while they are being examined by the 'grey' type of alien. The reptile form is supposed to behave subserviently to the other types of alien and may be of a trained lower species – an animal or equivalent.

The 'monster' form has been described by frightened witnesses as having a one-piece space suit and two glowing orbs where eyes would normally be in a human. Sometimes these entities give chase to witnesses and it is said that *they* do not appear to walk but rather hop, perhaps as an astronaut would on the Moon. Foul sulphurous odours are also associated with this entity.

GOBLINS

Goblin-type beings have also been seen, and perhaps the most famous case concerns the Kelly-Hopkinsville event of August 1955. A witness in a Kentucky farmhouse remembers seeing a huge shining object land close to the farm at about 19h00. Less than an hour later the farm dog started to bark furiously and the occupants of the farmhouse were astounded to see a small, shining 'man' with enormous eyes approaching the house with his arms raised high above his head. The occupants, being country folk, tended to shoot first and ask questions later (the decision no doubt being influenced by the extremely strange appearance of the entity). Both rifle and shotgun were used and it was said that when the entity was hit it sounded just like the noise you would hear from a shot into a pail.

The one creature was joined by others and *they* seemed to be able to float, so even though *they* were hit with pellets and bullets, this did not stop *them* from getting on to the roof of the farmhouse. The siege carried on for about three hours, until the occupants felt they had to escape to go and fetch the local sheriff. The Chief of Police, Russell Greenwell, Deputy George Batts, Sergeant Malcolm Pritchett and three other officers from the Kentucky State Police, plus a photographer from the *Kentucky New Era*, returned with the family to their farmhouse with the intention of taking up the fight with the intruders, but all they found were bullet holes. The intruders had gone. The policemen therefore returned to their station, but when the family had at last settled down for the night, one of the creatures was seen to be peering in through a window at a female family member. When fired upon, the creature disappeared into the darkness.

The hysteria following this unusual case was dismissed by some observers as religious fervour, as the family were said to be religious fundamentalists. However, it is significant to note that they never retracted or amended their version of the events that night and, while Project Blue Book never officially investigated the case, it remains classified as 'unidentified'.

R O B O T S

One of the most famous cases regarding 'robot'-type entities concerns the strange and terrifying ordeal of Robert Taylor, a Scottish forester. On 9 November 1979 he was inspecting the plantations near the M8 Edinburgh–Glasgow motorway, when on reaching a clearing he was confronted by a large dome-shaped object about 6m (20ft) in diameter, with what appeared to be a rim about halfway down the dome. He remembers that the object seemed to be phasing in and out of reality, becoming alternately semi-transparent and then solid. He suddenly became aware that two spiked, ball-like objects, each about 30cm (1ft) in diameter, were rushing towards him, apparently from the larger object. These objects grabbed him by the trouser legs and tried to drag him to the main object. He collapsed, being overcome by a choking, noxious smell. On coming round, he believes he heard a swishing sound but could not see any of the objects. His dog was racing around him, barking excitedly. Taylor struggled to his feet and tried to leave the area in his pick-up truck, but unfortunately managed to get it stuck in soft ground. Consequently, he had to walk about 1.5km (1 mile) back to his home. He later suffered a headache which lasted for some hours and a thirst which lasted for around two days or so.

This case is puzzling inasmuch as the incident occurred in daylight, around 11h00, near a major motorway, but no witnesses have been able to corroborate Robert Taylor's story. However, there were traces to show where the smaller objects had rolled over the ground but no burn marks or any other lasting signs. It is not known whether any radiation tests were carried out on the surrounding soil, but Robert Taylor remains convinced that he was the subject of an unsuccessful abduction attempt by some form of robot device. The fact that no other person on the motorway saw anything strange that day might be explained by the objects' apparent ability to become invisible.

G R E Y S

The 'greys' are by far the largest group of alien types in evidence. It is this group which appears to undertake most research on us and which consequently has the most contact with human beings. Unfortunately, people who have been in contact with the 'grey' entity mostly remember fear and apprehension. This group seem to be the most 'alien' of the visitors because *they* appear to have little idea of what makes us 'tick'. Their total disregard for our emotional welfare leads one to believe that *they* do not want to think beyond their immediate task: a physical examination, together with the taking of tissue samples or body fluids. Their practice of planting devices in our temporal lobes, ears or other suitable body places for unknown purposes also provokes profound fear and a feeling of helplessness.

Experiencers have often said that the greys seem to have a kind of hive mentality in their method of operation – as though *they* are being directed by some higher authority and are merely there to 'pull levers', as it were. One of the more frightening aspects of the greys is their apparent ability to manipulate the human mind so as to cause altered states of consciousness or amnesic episodes in their abducted subjects. Their ability to manipulate our minds is equalled by their reported ability to manipulate matter: the floating of abductees through solid walls or windows while at the same time 'switching off' a sleeping partner, or taking just one person out of a car occupied by several others without affecting people *they* are not interested in.

Such feats of technical superiority may or may not be a literal manifestation of alien science. It is possible that our knowledge of matter, and perhaps more particularly of our own physical and spiritual constitution, may be totally erroneous or just misunderstood. It could therefore be that in some cases it is not the physical that is being abducted but the ephemeral – the soul or human essence, if you like. The situation is further confused, however, by the fact that some people abducted by these grey aliens retain physical traces of their experience, such as wet clothes from being exposed to dew.

The taller, skinny greys seem to be fewer in number than the short greys. These taller types have often been seen orchestrating events when the examination of an abductee is taking place. *They* are frequently described as female, although it is not clear that *they* possess any identifiable gender attributes, which is true of the greys in general. The tall ones are often described as exuding a kind of love for us – perhaps it is the same kind of love a veterinary surgeon might have for a sick animal in their care? This kind of love calms the subject, so its purpose may be to make the necessary procedures less traumatic for the victim and less troublesome for the perpetrator. Or is it more than that – who can really tell?

The tall ones appear to take no part in the actual physical examination of their unwilling human subjects: the short greys do that. The tall ones seem to 'float' around the proceedings in an almost detached, voyeuristic manner. As this analysis is based upon the general impressions of probably terrified abductees, it may well be that the observation is somewhat less than objective and could be influenced by mind-altering techniques, the use of specific chemicals or by some other unknown but equally effective means.

There are, of course, many more types of reported alien, ranging from an amorphous jellyfish kind of blob to a 'being' of light. However, these types are not seen as often as the small 'greys', 'tall greys' and 'Nordics'. It might perhaps be thought unusual that more robot-type creatures are not observed, especially as we humans would no doubt make extensive use of robots to do our dangerous and strenuous work on other worlds, especially if our expeditionary numbers were strictly limited. It is worth remembering that the greys and many of the other aliens breathe our air and seem perfectly comfortable in our gravity, with no perceived problems. This may have come about through adaptation, technology or a programme of their own genetic modification. It could also mean, of course, that their world is an oxygen world like ours.

'Grey' physiology

The greys appear to represent the largest groups of aliens interacting with the human race at the present time, and later in the book it will be established that *they* probably deserve the most attention from us because of their actions.

The remarkable Whitley Strieber, author and experiencer, describes the autopsy of two small greys in his cryptic novel *Majestic* (see Bibliography). The account in the novel concerns the fate of two *apparent* (my emphasis) alien creatures, presumably as a direct result of the Roswell incident. One of the bodies appears to have been of a human derivative, possibly undergoing some kind of development to the 'mature grey' which is described next. The other is somewhat more radical and is clearly non-human in origin: such features as a multi-chambered heart, two distinct brains separated by a thin cartilage and an atrophied digestive system make this creature definitely non-human! There is no sexual means of reproducing the species (or no requirement for one), with neither male nor female genitalia in evidence.

The Whitley Strieber novel is just that – a novel. However, is it too great a leap to imagine that Strieber may have based his fictional narrative on first-hand witness accounts of the actual events that happened at Roswell, and later as a direct result of it? If the autopsy procedures and discoveries described are to be considered in isolation with not even some tenuous support, or possibly as just a partially substantiated rumour, the description of the events would carry little or no weight at all. However, some claim that corroborative but anecdotal 'evidence' does exist which lends support to the description of the creature's declared physiology.

The fact that the short greys have no perceived method of self-duplication means that either *they* are hermaphrodites (not likely, as not only do *they* possess no sexual organs, but *they* also possess no means of birth as we know it), or *they* are the product of some kind of genetic cloning method currently unknown to us. The biological reproductive systems of Earth seem to be of continual undying interest to *them* – particularly that of humans.

The cloning conjecture may not be so far out as it at first seems, even if we do not understand a mechanism of such sophistication on such a grand scale, particularly as the short greys exhibit the so-called 'hive' mentality and appear to be uniform – like replicas of each other. However, it is more likely that their ability to react in a strictly coordinated way, much as shoals of fish might react to a perceived threat, may be the result of a telepathic response rather than one mind at work. Perhaps their minds (or part of their minds – their speech function appears to be as atrophied as their digestive system) can be likened to radio transceivers which are always switched 'on' or on open channel, with a multitude of channels available. If this were the case, it follows that *they* could all receive a message at the same time and if any one entity replied it would be heard by all. The advantage for the greys would be an elimination of possible misunderstandings arising from garbled messages or instructions – especially important when it is necessary for crew members to act in a critically coordinated way when travelling across space or jumping time. From a classically human perspective, the disadvantages of such a system would be cacophony and disorder, because of our instinct to act alone and only to function as a team when trained to do so, but it is easy to see that if the system were well regulated it could work very effectively indeed. Our perception of such a system might well create the impression of a 'hive'-like response. Overall, it is probably realistic to consider both clone and transceiver hypotheses as equally valid and not mutually exclusive.

If the Roswell story is true, it would appear that the US government has known about the physiology of the short greys at least since 1947, when several alien cadavers became available and were reputedly autopsied. However, the claim by some researchers that autopsies *were* carried out and that one alien survived the crash and was held by the US government has not been substantiated, even though there is a lot of circumstantial evidence that the event actually did occur. If the story is largely true and not a cover-up born of embarrassment over Project Mogul, then no doubt much more is already known in some quarters about these particular visitors than simply their physiology and their means of communication – telepathic or otherwise.

ALTERED STATES OF CONSCIOUSNESS

Gurnemanz's reply to Parsifal's question 'Who is the Grail?'

That tell we not; but if thou hast of Him be bidden
From thee the truth will not stay hidden.
The land to Him no path leads through,
And search but severs from Him wider
When He Himself is not the guider.

Parsifal
Richard Wagner

I n 1987, a remarkable book entitled *Communion* was published (see Bibliography). The author was Whitley Strieber and the book told of his struggle to come to terms with an incredible assault by the 'grey' aliens into his life.

Strieber initially recounts his feelings of absolute helplessness in the face of the greys and an almost overpowering anger that *they* can do what *they* like, when *they* like, not only to him, but also to his wife and son. The story of *Communion* and beyond (*Transformation – The Breakthrough* – see Bibliography) is remarkable because it tells of serial abductions going all the way back to his childhood. Mr Strieber is a successful novelist who had several books to his credit before he published the ground-breaking *Communion*, and after his strange experiences with these non-human entities he speculated that he was somehow prepared or coached by them for his series of contacts.

THE SPIRITUAL DIMENSION

As with others who have experienced a powerful and prolonged contact with the greys, there is in Strieber's account a very strong sensation that something spiritual is happening. Somehow, the experiencer comes out of an exchange with the entities – even though that exchange may involve what amounts to rape – with a heightened spiritual awareness. It seems a peculiar exchange: the loan of one's body for an increased spiritual awareness. Sometimes victims describe a distinct tendency towards clairvoyance and enlightenment following an encounter.

The actual manifestation appears, superficially at least, to be equivalent to the popular transcendental experimentation of the 1960s, where hallucinogenic drugs were used to achieve a sense of altered reality and a creative state of mind. Initiates believed that they were experiencing real contact with God or the spiritual 'whole'. Interestingly, sex was also used as an expressive instrument in the initiation of the search for 'self', as it has been since time immemorial. However, the use of powerful mind-altering drugs to achieve that state should not be considered similar to the methods the abductors use, unless the techniques of alien abduction cause a similar response through the non-chemical manipulation of the mind. In fact, the abduction can create an anoesis in the subject, with the loss of normal independent action or power of thought – generally unlike the effects of mind-altering drugs. The greys seem able to invoke much more powerful and lasting responses, which hypnotherapy may only partially bring to the conscious surface in patchy episodes of clear memory recall.

Some abductees are said to experience a feeling that they are renascent, or a 'better person', but most soon find that the experience tends to isolate them and create unusual problems which not only affect, and sometimes even destroy, their lives, their work and their family relationships, but also cause them much inner turmoil due to the frustration of not really knowing anything about their abductors or the reason why they and not someone else were taken. There is also the terrible fear that *they* may come to take the abductee again at any time – night or day. These feelings of fear, guilt and rejection will, hopefully, result in psychiatric assistance being given to ease the subject's burden so that they can get on with their lives. Unfortunately, sometimes the burden is too great and no help is forthcoming or the symptoms are not recognized. In that event, the victim may not survive.

During the course of his various experiences with the 'visitors', Whitley Strieber was shown apocalyptic scenes of the destruction of Earth. How this was achieved is not known. Whether it was a mind trick induced by unknown technology to see what kind of reaction such scenes would create in their victim's mind is open to question. Whether it was a view of our inescapable future is also unknown.

Apocalyptic visions of the destruction of Earth appear to be commonplace with abductees and seem to be a central plank of the greys' abduction repertoire. Perhaps they are staged to awe the abductee and get them in the right frame of mind. In any event, the procedure appears to achieve the desired result, with the abductee very much sobered after the experience.

Students of the subject will notice the somewhat strange progressive nature of the abduction-by-aliens experience. This is much the same as the way in which the so-called 'cultural tracking' of aerial sightings seems to have dominated our perceptions over the past ninety-five years. The objects have ranged from airships near the turn of the century, through 'ghost' rockets in the 1940s, mother ships with their small reconnaissance saucers in the 1950s and 1960s, and now back to the almost archetypal flying disc. In the first abductions, it appears that the visitors were interested mainly in our physiology and how we reproduced ourselves. Having examined their unwilling subjects, *they* let them go. Now, some female abductees report repeated egg-taking, as well as the unwilling breeding of hybrid entities taken from their wombs before term (whatever that may be), to be bred on by the aliens themselves once a particular point has been reached in their embryonic development. This seems to indicate that the aliens are either very interested in an experiment to see if human females can successfully carry a hybrid entity in their womb, or that currently *they* are unable to carry out the whole process themselves. It also strongly suggests an alien timetable.

Present-day male abductees, while being continually raided for sperm samples, give an equally disturbing account of other aspects of their abductions. It would appear that there is a degree of 'incorporation' going on. 'Incorporation' describes the act of an alien entity taking over a human body, and once this has happened it appears that the original human personality ceases to exist. While the proposition seems utterly crazy, it should be remembered that many people regard reincarnation as a fact and there are numerous accounts of inexplicable events occurring with no alien dimension at all.

A constant and recurring theme with present-day abductees is their feeling of empathy with the visitors. No longer are the aliens generally seen as nightmarish fiends, intent upon some nefarious activity akin to rape, but rather as teachers, as ambassadors of creation and the Universe. The purpose of the act of abduction is merely to gain our individual attention so that the necessary messages of salvation may be given to those chosen to carry them.

In a work such as this it is sometimes necessary to play the devil's advocate. So, the whole process could, of course, be a figment of a number of diseased imaginations. The need for chastisement and punishment for sins, real or otherwise, runs deep in the human psyche, and it seems necessary for a large section of Western society in particular to feel guilty about the many advantages they have. (The remaining population probably do not feel guilty because they are too busy *being* guilty!) Perhaps that is why this 'new religion' of brothers from space is gaining ground over existing organized religions, which can offer little more than tired dogma with little relevance to today's world.

However, the growth of stories of hybrid breeding and 'incorporations' may mean something more than a simple masochistic disorder. While some accounts are no doubt the result of 'copycat' responses, enough cases exist worldwide to make the problem a real one. Somewhat predictably, the only people currently exposed professionally are therapists, psychiatrists and psychologists who look at the situation from a case-study point of view, usually against a backdrop of conventional psychotherapeutic techniques with an overriding commitment to the patient, whom they wish to 'cure' in order to make their life bearable on a day-to-day basis. Governments do not express any open interest – but how can they, since they do not openly admit to the visitors' presence?

SPREADING THE MESSAGE

In almost all abductions, the subject is given a message to spread to the rest of humankind about the error of our ways, and the dire warning that if we do not change the human race will suffer extinction. The message of extinction, or at the least very nasty consequences indeed, is nearly always the same, ranging from nuclear wars that will wipe us all out to huge natural disasters (similar to some of Nostradamus's predictions), and is always on a megalithic scale of

'One' – it is never of a minor nature. Whitley Strieber claims he was told of the dangers of the ozone hole over the Antarctic in February and March of 1986, long before scientists understood the significance of the discovery themselves. He prophesied that there would be measurable crop damage from excessive ultraviolet light in the years 1990–93. He also claimed that atmospheric problems would cause disease in all animals (humans as well?) due to their immune systems being affected by the excessive ultraviolet light, although he admitted that he did not fully understand the mechanism of how that might evolve. He also claimed that there would be some reduction in the ozone holes from 1987–97, but the respite would be only temporary.

It is difficult to see how Whitley Strieber hoped to do anything with this information, because unless it were linked to a remedy, a course of action humankind could take to avert a potential or impending disaster, it would be useless to us. With all due respect to him, it seems unlikely that world governments would take much notice of someone who essentially earned his living by writing horror stories! Perhaps the intelligence behind the statement just wanted to make us feel bad about ourselves? Perhaps Whitley Strieber should have asked his abductors to elaborate? It now appears that the issue of global survival lies in the hands and depends upon the actions of various 'green' movements who, at best, physically intervene to reverse events that may harm the environment. The victory of Greenpeace in 1996 over the deep sinking of a defunct oil platform at sea proves a point, even though the outcome of that victory may not have had the desired effect – at least the green movements raise the general public's awareness of environmental issues. So, are the aliens' messages really superfluous? Or, on the other hand, is it that the green enthusiasts have heard the message and are the only ones who are actually doing anything about it? Or, is someone playing games?

One of the most puzzling aspects of scenarios where messages are given to an abductee is the question of why the aliens don't select someone who is credible and professionally qualified to understand and act on their warnings. In the Strieber case, perhaps they should have selected a meteorologist or a biologist, or even both, to carry the message about the ozone hole. Surely we would take a warning about impending atmospheric doom more seriously from someone who had a greater knowledge of the subject to start with and who could probably weave the message into their area of expertise? In that way, they could call professional bodies together to discuss the issue, and even perhaps address the United Nations (UN) with their findings. The message would have far more credibility if it came from a scientist who had a grounding in the subject, rather than someone who appears as a lone prophet in the wilderness, despite having the prophet's conviction.

There are, of course, several possible explanations as to why scientists and their like are not selected. There does appear to be a trend of using serial or favourite abductees for the steady release of alien messages or pronouncements. If the aliens gave the power of being forewarned of these potential or impending ecological disasters to a scientist, there would be a real risk that they simply would not believe it. If they did, it would have a serious impact on their scientific paradigmatical training – and who would believe them anyway? There is also the possibility that they might use the information for their own self-aggrandisement... There seems to be some logic in the aliens' selection method after all.

We have had warnings of global human destruction by nuclear war, by natural disaster and now by a depleted atmosphere. If the 'cultural tracking' system runs true, we should shortly be warned of the dangers of genetic engineering. Clearly, if nature had intended genes, animal or vegetable, to be transportable across species, she would have arranged it so without the need for human intervention. Evolution on Earth relies on active species involvement, where two sexes come together to form a species progeny. In lower animals and plants, the process creates an exact image of the parents to ensure species survival. Occasionally, a new species or variant evolves. Nature tests evolution all the time: the weak or sickly animal or plant does not survive and therefore a species can disappear as a natural corollary of nature's experiments with the occasional appearance of a new prototype or mutation. In the higher animals and human beings, such potent destructive evolutionary processes seem out of place in our strangely compassionate societies in the West, which seek to frustrate evolution at every turn while at the same time making better weapons with which to kill and maim those who would threaten a way of life.

The aliens may have found their own solution. Our model of evolution on Earth cannot be the only model which exists, and it is quite possible that our visitors have overcome the 'two opposite sexes per species' model, thereby taking charge of their own evolutionary processes. If that is the case, we may be very low on the evolutionary scale indeed.

ASPECTS OF ABDUCTION

Many modern abductees seem to get more out of the experience than their earlier counterparts. Quite how that works is open to question, but essentially the whole process seems to be telepathic. It does not seem to matter what the nationality or language of the abductee is – the aliens are able to speak to them telepathically. Even more importantly, the abductee also gives the appearance of being able to transmit *their thoughts to the entities*, so that a conversation takes place in the respective minds – a feat which, naturally, is outside the subject's repertoire in the normal course of their life. It could be, of course, that no telepathic process is being initiated by the subject at all and their mind is being read by the entity and questions anticipated by an unknown mechanism. If that is the case, humankind has no defence against the aliens' mind-scanning techniques.

More and more abductees report quite detailed events in their experience with the aliens. They give accounts of being broken down molecularly for transportation purposes ('Billy' Meier – see page 18 – apparently experienced this mode of travel on occasion: could it be similar to the *Star Trek* transporter?). They also report details of their past lives on other planets and star systems, which seem to support much of the Eastern doctrine of reincarnation. Some remember earlier deaths on other worlds.

Normal past-life experiences (if one can consider a past-life memory normal at all) are always concerned with a life on Earth, not usually on another planet, and sometimes relate specifically to a period in Earth history when great apocalyptic events occurred. Such traumas for humankind occur in events like the Black Death in Europe, or the French and Russian Revolutions, or maybe natural geological disasters.

It is perhaps amusing to note that some 'conventional' past-life narratives include stories of being an Egyptian handmaiden or maybe a eunuch in an Arabian court. Sometimes a female past-life experiencer will claim to have been one of the Queen of Sheba's handmaidens, or a man may claim to have been a knight in medieval Europe. Many of the stories that are told no doubt relate to wish fulfilment rather than actual events.

However, if the above examples are considered weird and outlandish, how much more so if you claim you came from another planet and a star system light years away from Earth? The peculiar thing is that most past-life experiencers seem to be able to describe their medieval or pre-medieval lives in minute detail, and the 'other planet' person can match these with equally plausible descriptions of their other lives – albeit much stranger than even the strangest past life on Earth, but, of course, quite impossible to verify in any respect. It is also interesting and, some would say, predictable, that the 'other planet' people are all abductees.

It is apparently rare for 'conventional past lifers' to be also 'other planet' people. It is not known whether this is because the aliens only abduct 'other planet' people by some means (a monitoring system across time, perhaps?), or because an abductee is conditioned to believe that they have had an 'other planet' life as a result of the abduction itself. If that is the case, a possible mechanism may be a powerful but conventional psychological reinforcement technique enacted on the human subject to make them pliable and suggestive for future encounters.

The removal of aggression from human subjects before an abduction seems to be a very important consideration for the alien greys, in particular, who seem to fear any physical response to their actions. If it were a question of weight against weight there would be no contest: a grey may weigh only 23kg (50lb) or so, compared to an average human male of around 73kg (160lb). It is natural, therefore, that if you wanted to examine, somewhat intimately, another being three times your weight and with muscles you had only dreamed of, you would want your subject sedated as much as reasonably possible! It is probable that *they* are also responsible for taking cattle, sheep and horses, and it is interesting to note that *they* appear to have no fear of large animals, which can weigh up to 450kg (1,000lb) or more. (It should be mentioned that no large animal is on record as having been returned alive – unlike human subjects.) All in all, it would seem that the greys do not want to upset us too much, otherwise we may respond in a way which may be regrettable for both sides.

Students of the subject will know that it is alleged that armed responses have occurred in the past, usually involving fighter aircraft on intercepts. Such encounters usually turn out to be somewhat innocuous affairs, due to the fact that aircraft on-board weapon systems are neutralized before they can be activated. However, there have been occasions where aircraft have been downed by discs, although the celebrated but unfortunate case of Captain Mantell is now widely thought to be an instance of pilot error (but see page 32). There have also been alleged cases of contacts with armed military personnel where greys have been shot by small arms fire. However, these events would only be recorded as the result of gossip or hearsay – they would hardly be admitted by governments or base commanders.

The spirituality facet of the encounter is an interesting one to explore. It could mean that the greys know more about us than we know ourselves and have the means to tap into not only our intellect, but also our soul. *They* appear to be able to give the victim precognitive skills and a heightened perception of the Universe, together with a degree of clairvoyance where none existed before the encounter. How, or why, that should be is unknown. It could mean that there is a genuine exchange between us and *them*, in order to fulfil an agenda of gradual human enlightenment not only as to their presence, but also as to our place in the Universe as a whole. However, this thesis may be flawed.

The exchange of one's soul for knowledge is, of course, a myth which persists throughout the history of the world. Could it be that our ancestors, struggling to explain bizarre events and happenings beyond their experience and imagination, described this process of seduction and exchange in the context of their religious faith as the work of the devil? If that is the case, the process of abduction has been going on for a very long time indeed. Less developed religions may have tended to view such encounters as being with nature gods or elemental spirits, sent to torment humankind as the result of displeasure and subsequent punishment.

The question of hybridization is an increasing concern, not simply because it is claimed to be going on, but because the greys seem to have been able to produce their own hybridized humans in order that these hybrids may live and work among us without being detected. The implications are enormously important. There seems to be no way in which these hybrid creatures can be unmasked and *they* reportedly have influential jobs in high places. However, it should be remembered that these kinds of frightening claims originate from the fringes of ufology, which may have their own agendas.

ABDUCTEE VERSUS CONTACTEE

A distinction needs to be drawn between the abductee and the contactee. Generally, the abductee has no choice in the matter. The contactee, however, appears to be a rather different individual who is favoured by the aliens, who return repeatedly to the subject, perhaps in much the same way as the mythical vampires go back to their prey time after time. Why this should be is not immediately apparent, except in cases where there appear to be genetic implications, i.e. a possible breeding or hybridization programme. Such programmes are almost universally ascribed to female abductees.

Contactee males appear to be given a different mission, in that they are instructed to go forth and give the world the message of peace and goodwill. There is no point in the aliens telling this to the holders of power, for they would never believe it. In any case, the aliens might have another programme for them. The contactee scenario has little to commend it, as it will later become clear that the aliens have no interest in making friends of anyone and appear to treat us as we might treat a moderately intelligent animal. This is especially true in the case of those humans who have no particular role to play other than to be pawns in the extremely strange alien game. It may well be that the so-called contactees are either serial abductees who *imagine* that they have a special purpose, or abductees have had just one interaction and yearn for more, or individuals who possess a latent desire for fame, fortune, even notoriety, and see the whole alien issue as a means of achieving their aims. There may be others who steer even weirder courses.

A 'strange' case

A particularly strange case involved a Dr Frank E. Stranges (no pun intended), who claimed that he met and talked with a resident of Venus called 'Valiant Thor'. A few words about Dr Stranges. He is presented as a Founder and President of the American 'National Investigations Committee on UFOs'. He is also President of International Evangelism Crusades (a worldwide Christian denomination) and the International Theological Seminary of California. He holds degrees in Theology, Psychology and Criminology. Dr Stranges is also an Assistant Deputy Director of the California State Marshals' Association and a Chaplain of the American Federation of Police. He has investigated the UFO phenomenon for the past forty-four years. In Dr Stranges' booklet *Stranger at the Pentagon* (see Bibliography),

Harley Andrew Byrd (nephew to the late Rear Admiral Richard R. E. Byrd, US Navy) claims that Project Blue Book (in which he was a participant) received an urgent message in mid-March of 1957 from the Alexandria Police Department, stating that two police officers had picked up an alien who had landed about 22km (14 miles) from Pentagon Boulevard. It is claimed that the alien was taken to the Pentagon to meet the Under Secretary of Defense, and then on to President Eisenhower and Vice President Richard Nixon. It is said that the meeting lasted around an hour or so before the alien was returned to the Pentagon.

The 'space emissary' (as labelled by the Department of Defense) supposedly met with the Secretary of Defense, D. F. Forestall, and others who numbered twelve (Majestic 12?). Dr Stranges, who had been speaking at the National Evangelistic Centre, was then introduced to Valiant Thor. It was said that Valiant Thor stayed on Earth until 16 March 1960 (three years), at which time he disembarked for home – Venus. He indicated that not only did his race live underground on Venus, but many of the planets in our solar system also harbour sentient life in the same manner. Before he left he spoke of Christ's presence in the Universe, and that it was heart-warming to see that His advanced teachings continue.

The series of contacts between Dr Stranges and Valiant Thor (who incidentally looked like an average late 1950s early 1960s dresser with a handsome, clean-cut appearance) unfolded in a somewhat unremarkable way considering the totally incredible scenario of an actual Earth landing by an alien. When meeting with President Eisenhower for the first time (Valiant Thor did not only look totally human, he could also speak the language), Eisenhower is supposed to have asked him for proof that he was who he said he was. Thor is said to have asked Eisenhower to go with him to his ship. Eiesenhower naturally demurred, saying that he was not free to go where he pleased. Richard Nixon then burst in on the meeting. He immediately rushed over to Thor with outstretched hand. Nixon is supposed to have told Thor that he had created 'quite a stir... for an out-of-towner'.

Valiant Thor apparently regularly 'beamed' back to his ship and friends, Donn, Jill and Tania (crew members), in the three years he was on Earth. In his professional role as a criminologist, Dr Stranges was intrigued at Thor's lack of fingerprints. According to Thor, humankind had been marked with fingerprints since the fall of Adam in the Garden of Eden, and increasing theological input from Thor convinced Dr Stranges that Christ is the rightful ruler of the Universe. Valiant Thor's mission on Earth was supposedly to rid it of disease and to enlighten us as to the perfect way of life.

Plans were made to obtain a Presidential address to the UN, which was rejected by the Central Intelligence Agency (CIA), Secretary of Defense and the military chiefs of staff. A press release to the UN by way of the Secretary General was also promised no later than 7 February 1966. However, despite attempts by Nixon to get Eisenhower to make the decision to let the world know of the plans of Valiant Thor, the process was blocked by bureaucrats.

After Valiant Thor had left Earth, Dr Stranges continued to have periodic meetings with him, as well as being in 'holographic communication' (like Princess Leia in *Star Wars*?). Dr Stranges recalls, with excitement, how he was taken aboard Thor's craft. He was ushered to a special room and asked to disrobe and walk through a purifying compartment. After being given a special one-piece suit, he was taken into another room which turned out to be Thor's quarters. When Thor pushed a button an entire wall became transparent. He told Dr Stranges that he had met with Bobby Kennedy and had attempted to dissuade him from standing for the presidency for four years. However, Kennedy ignored the advice, which Thor repeated several times. We all know what happened next.

Dr Stranges was shown to his own guest quarters. The lighting was pale blue with no direct source of illumination. The flooring was moulded and fitted to one's feet and everything was hidden in the walls of the room. Dr Stranges recalls how he went to the bathroom and was embarrassed to learn that there was no toilet tissue. However, he was saved by the voice of Valiant Thor in his head, telling him to push three wall buttons in descending order. On doing so, warm air blew beneath the toilet seat. The effect was to crystallize the 'waste matters', causing them to drop from him. The second button produced a not dissimilar effect, while the third made Dr Stranges feel as though he had been 'washed, cleaned, powdered and perfumed'.

Dr Stranges was so impressed by the experience of actually being on Valiant Thor's craft that he outlined the general layout of the craft which included a kitchen, a cafeteria and a dining area, together with sleeping quarters, maintenance area, vehicles, stores, laboratories and, of course, the all-important power plant.

I suppose one must call Dr Stranges a contactee of the era when people were relaying the message of the 'space brothers', who were coming to help the poor human race redeem itself from its own iniquitous and corrupt vileness. Really? Are we *that* bad, or is it simply a basic masochistic need of humankind to be so chastised all the time? Do we need this collective push of disapproval to do better, much as a child who can do maths but is too lazy to attempt the subject? Dr Stranges' somewhat peculiar (and possibly unique) mix of religious zeal with a pioneering nature may be performing the function. However, the time-honoured core message of goodwill to all men may here have become too entangled with the complexities of the visitor experience itself, and some readers may perhaps feel more comfortable in dismissing the whole affair as irrelevant to the real quest for truth.

OBTAINING EVIDENCE

Mention should be made of the manner in which much abductee evidence is brought to light. It must be remembered that many of the abductees will remember events over a period of time because of the natural process of memory recall. It is perhaps significant that the techniques employed by the aliens, despite their obvious technological superiority, cannot completely erase memory while leaving the brain intact and fully functioning (this legitimately presupposes that *they* cannot erase memory totally without permanently damaging their subject or their implanted devices).

Regression hypnosis is a tool which is often used by psychoanalysts to bring a patient's suppressed memories to the surface. Serial abductees (not contactees), while having a greater understanding of the actual trauma, may still have quite a complicated blocking of memories, compounded by the number of times they have been abducted. In fact, some people may never know that they have been taken by aliens, except that they may have an irrational fear of a certain stretch of road or of a particular place. They may also have unusual marks on their bodies which they cannot explain. Such unfortunates might just be classed as neurotic individuals and society may simply leave it at that, unless the people concerned are motivated or encouraged to take matters further.

While hypnosis is an excellent tool for uncovering suppressed memories, it should be remembered that it is not infallible. The Achilles' heel of hypnosis concerns the ability of a subject to relate an experience that they truly *believe* they have had. All that matters to the hypnotized subject is that they retain a commitment to their belief. They may also communicate the misjudgements, hyperbole or observational errors that they *believe* they experienced as, under hypnosis, they are *telling the truth as they saw or experienced it*. It therefore becomes clear that an actual objective event which may or may not have happened is, somewhat strangely, secondary to the experiencer's *belief* that it actually happened. It must also be remembered that something unnatural has happened to an abductee which is, by definition, completely outside *anyone's* normal terms of reference, so it is reasonable to allow for some distortion in the telling.

It is an interesting fact that most abductees tell of an implicit sexual interest in them on the part of their abductors. A classical Freudian psychiatrist would find many normal reasons as to why individuals lean towards such sexual exploration. However, if the events are objectively real, there could be reason to suppose that the *only* interest the aliens have in us is connected with our ability to reproduce ourselves.

The imperfection of hypnosis as a tool for assisting in abduction cases is not diminished by the problem of literal recall, because it is known that some distortion may occur and allowances can be made. This is particularly true if more than one individual has had, or seems to have had, the same experience, because individual recalls can be checked against one another. However, a greater risk of error in uncovering the objective truth may occur because of the actual methods of hypnosis. It is very easy to lead, or implant suggestions in, a subject undergoing hypnosis, and if untrained therapists are careless in their approach it is possible to contrive results which may be at complete variance with the objective truth – whatever that is.

The ephemeral nature of the subject of abduction is not helped by the fact that there is generally no corroborative evidence. It may well be that the experience of a 'past life' or an abduction is related to a neural phenomena called ' cryptamnesia', where memory recall is drawn from long-forgotten memory. It is very probable that the complexities of the mind filter, mix, sort and polish experiences either on their own, or as a direct result of suggestions innocently inferred by the interrogator. The *modus operandi* of abduction is usually one of stealth and the experiencer is left disoriented, lost and totally confused at the end of it. Untreated, they can suffer both physically and mentally. In extreme cases, it is possible that psychosis may result, with all the risks involved with severe depression. The depression inevitably deepens when the abductee decides to talk about their experience, because it usually becomes an act of faith to get the listener to believe their story, even if they are a close relative.

It is unrealistic to think that all abduction stories are objective truths. The reasons why someone may want another person, or the public at large, to believe they have been abducted by aliens when they have not are legion, not least being the question of gain. Deliberate hoax stories are, however, outside the scope of this book and rightly deserve the attention of the psychiatric profession.

Even so, there are an increasing number of people coming forward claiming that they have experienced bizarre episodes and interactions with alien entities. Whether the entities in question are *our* aliens or whether *they* are discarnate beings, is a matter for the specialist analyst who may be able to differentiate between the two (if that is possible). Nevertheless, some analysts have written up their research findings, with the permission of their subjects, and the accounts often provide a fascinating insight into humanity struggling against something which it can barely even begin to understand.

A recurring theme is that of serial egg or sperm theft, and possibly *in vitro* (IV) fertilization, together with the repeated removal of immature foetuses from the host 'mother' over a protracted period. This recurring theme superficially indicates a classic Freudian scenario and would quickly be dismissed as just that, were it not for similarities in the base lines of many of the stories told by unconnected people. Despite these similarities, however, it is right to consider the possibility that some witness accounts do not emanate from an actual abduction event and are rooted in a deep desire in the subject to be noticed and taken account of as a person by their peers and society at large – after all, there are enough books and films around now which publicize the abduction scenario well. However, from the analyst's point of view even those cases should be treated as valid, *real* events until proven otherwise, for as far as the subject is concerned their experience is real to them, and requires expurgating from their subconscious in order for them to return to a state in which they may lead a reasonably normal life without the baggage of a crippling neurosis.

QUESTIONS OF CONSCIOUSNESS

John E. Mack MD, Professor of Psychiatry at Harvard Medical School, has published an account of thirteen cases of possible abduction by alien entities, *Abduction – Human Encounters With Aliens* (see Bibliography), in which the phenomenon is presented both succinctly and with great respect for the subject. It seems that his work has caused some bad feeling among his peers at Harvard, who are very upset that he has dared to suggest that the phenomenon of abduction is real and not the product of psychotic imagery. Professor Mack (J.M.) perceptively reminds those critics who propound the 'false memory' syndrome as a blanket denial of alien abduction that witnesses have a peculiar lack of suggestibility under hypnosis which tends to substantiate, rather than diminish, the story: that is, when J.M. suggests to his hypnotized subject that corners exist in the room described or that the aliens have body hair, it is immediately denied – an important point.

Interestingly, J.M. also questions whether consciousness is the *only* method of knowing. Whether he is suggesting, as did Carl Jung, that there is a collective human psyche or archetype from which all our base memories come is not made clear. However, given the obvious superiority of the aliens over us in this area, it is reasonable to think that *they* may somehow be manipulating these memories in order to influence and control our reactions to *them* in a predictable way.

J.M. confirms that the support group method is a valuable adjunct to analysis. This is, of course, no different to any other support group therapy system and could equally be applied to alcohol or chemical dependency groups. However, there is a real risk of a non-objective psychosis becoming reinforced and deepened by the use of this method, so that the scenario actually becomes even more real to the subject than it originally was.

J.M.'s subjects recount familiar themes of world ecological disasters, apocalyptic visions of a splitting Earth and depopulation on a massive scale due to uncontrollable disease. The message is a continual and recurring one of dire warnings that if humankind continues on the same path it is on at present, such events *will* come to pass. Whether such terrible images and warnings are created by the aliens in order to stupefy and confuse the mind of the abductee is unknown. If archetypes do exist, it could be that the aliens may extract the fear archetype from the abductee's mind to assist *them* in the abduction.

As described earlier, the abductee sometimes comes out of the experience transformed, that is to say, more aware of life (on other planets as well as the more parochial view), of the world as a discrete entity and of the connectivity of the systems which support us. We have already seen that similar euphoric emotional responses are possible with the use of mind-altering drugs, but also that the classical responses using hypnotherapeutic drugs in a session will not have the longevity of an abduction experience.

This eschatological approach tends to undermine the possibility that the message is real and should be heeded. However, how can *one* ordinary individual influence world affairs, even if that individual is multiplied by as many as a million or so? If the aliens know all about us, *they* also know of our animal fear of death. *They* must know that human beings invent religious paradigms to help them live with that fear. Could it be that the aliens are merely exploiting their victims' belief system? They must also appreciate the risk in becoming the subject of a religion themselves. (It would be interesting to note whether or not the aliens' approach differs when presented with culturally disparate abductees or abductees with no religion or faith – does the person with faith become *more* religious after an abduction, or does the person with no faith *get* religion after an abduction? Also, is the effect the same when applied across different cultures?)

It seems, therefore, that visions of apocalyptic destruction might be an artefact to fulfil the abduction mission objectives and provide a degree of soft confusion to the victim after the event in order to cloak the *real* happening. In that way, the aliens may legitimately perceive that *they* are protecting their victim from the psychological damage which should naturally follow such a shattering event if the victim were mentally 'unprotected'. If that is the case, the aliens' system of blocking has failed both *them* and their victims.

Without getting into the possibility of there being different extant realities, an abduction almost always confuses the subject as to the demarcation between reality and non-reality, a conscious state and a non-conscious state. There is a blur between being awake and asleep, in much the same way that an anaesthetic functions just before one loses consciousness. However, abductees invariably report that examinations by their abductors occur while they are in a dream state, which may require the exorcism of psychoanalysis at a later date to return the subject to normal health.

Occasionally, the abductee is *shocked* out of their body. This is interesting, because most people cannot be *shocked* out of their bodies even with extreme provocation – it normally requires some skill and a degree of practice before ordinary people can accomplish it, and rarely does it happen spontaneously.

The question then is: 'Does the abductee *imagine* (through mind manipulation by the aliens) more of the objective abduction, or is the blurring of human reality a technological by-product or a necessary construct of the abduction itself as a means of controlling the subject? In any event, whether or not the victim has undergone an objective abduction experience, the accrued baggage of suppressed fears and anxieties will invariably lead to mental health problems at some time in their future – an aspect which is overlooked or ignored by the ruthless and all-powerful abductors.

HARDWARE AND OTHER MATTERS

A visiting angel to John Dee:

Ignorance was the nakedness wherewithal you were first tormented, and the first Plague that fell unto man was the want of science... the want of science hindreth you from knowledge of yourself

From *True and Faithful Relation*
John Dee, sixteenth-century English Magus

THE CASH/LANDRUM CASE

The night of 29 December 1980 found Betty Cash, her friend Vickie Landrum and grandson Colby driving towards Dayton, Texas. At about 9pm they had reached the suburb of Huffman, Houston, and they suddenly became aware of a very bright light, which rapidly descended to treetop level and remained stationary above the road about 42m (135ft) in front of them. Having got out of their car to watch, they noticed that flames appeared to be jetting out of the object. They soon decided that they would quit the scene, as they were all shaken by the sight of the thing. The flames seemed to emit a similar sound to that of a flame thrower (a roaring shriek) and they reported that a 'beeping' sound persisted throughout their encounter (Betty and Barney Hill also recalled a 'beeping' sound – see page 16). They followed the object, which had now moved off, and soon noticed that there was an escort of several Chinook-type helicopters maintaining station about 1.2km (¾ mile) away.

After dropping the others off, Betty Cash arrived home at about 9.50pm. She soon developed horrific symptoms which included terrible headaches, pains in her neck area and blister-type eruptions on her head which burst, letting out clear mucus. Her eyes became so swollen that she could not open them and she suffered from both nausea and diarrhoea. She was eventually admitted to hospital as a burns victim, but specialists failed to diagnose her ailments. Her hair started to fall out and she developed a bald patch. She eventually developed breast cancer and had a mastectomy. The others in the car developed similar but less dramatic symptoms, but all of them exhibited the classic results of being exposed to very large doses of radiation.

The witnesses decided to take the matter to law and attempted to sue the US government for considerable damages for the loss of their health and damage to their lives caused by the encounter. After years of legal wrangling and at a cost of millions of dollars, the case was eventually thrown out by the judiciary on the basis that neither the US government nor any US government agency owned the vehicle which caused their medical problems. Reports note that no emphasis was placed on the fact that the Chinook helicopters in pursuit (or escort?) were seen by other witnesses, and it was considered that the testimony of the expert witnesses in the case was a good enough reason to reject the Cash/Landrum submission.

Rumours abounded as to what the object they had seen actually was and ranged from a recovered alien craft being flown by human pilots, to an experimental nuclear-powered vehicle which had gone out of control. The Cash/Landrum lawsuit has become something of a landmark inasmuch as it demonstrates how difficult it is to challenge the authorities even when ample evidence of material damage and harm exists.

There is a peculiar synchronicity between this case and the Woodbridge incident in the UK (see page 20), in so far as two forestry workers maintain that the Rendlesham Forest incident occurred on the same day – 29 rather than 27 December 1980 – as confirmed by USAF Woodbridge Base staff. A coincidence?

It is reasonable to assume that if the US government owned such a device they would hardly let it roam wildly over inhabited areas, even if it was a remote-piloted vehicle (RPV); they would be even less likely to do so if it had a human pilot on board. In the former case, its flight could presumably have been terminated by its controllers, and in the latter the pilot could have undertaken an emergency landing well before it approached any populated areas, including highways. Some researchers claim that no evidence exists to rule out the possibility that the device was connected with the US government. Whether or not it was *owned* by them may be a crucial point and one in which the judiciary are correct. The observation that flames were seen jetting out of the object seems to suggest that Earth technology was at work, but it is not known whether the flames were paramount or incidental to the object's performance.

It is not known what type of radiation was responsible for the burns and other effects that the three witnesses suffered. However, it is possible that it consisted mainly of microwave radiation and not potentially lethal gamma emissions. The injuries sustained by the trio seem to point in that direction. This could also explain why no other injured witnesses came forward, as the microwave emissions could have been very localized or even directional. The device hovering in front of the victims' car is inconsistent with a vehicle in possible difficulties and more in line with a stimulated interest on the part of whoever was piloting it, or it may have been essential that the device was sighted for some reason as yet unknown. It now appears that the world will never know the real story of this remarkable but unfortunate incident.

The emission of microwave radiation from discs is not a rare occurrence and several witnesses have come forward stating that they have been burned by unseen jets or blasts of 'air'. The ignition systems of vehicles are regularly affected by the close approaches of discs. Domestic and industrial power supplies are also on record as having been affected by a disc's propulsion system. Nuclear missiles 'safe' in their silos have been neutralized electrically. Fighter aircraft which try to intercept discs encroaching into national air space have had their instrumentation and weapon systems neutralized. And yet domestic passenger-carrying aircraft appear to escape the ionizing effects of electrical disruption. Does this point to a disc's *weapons* system? If it does, it is a system which relies on a passive or negative reaction to a positive action. If it does not, it could mean that some discs radiate this nullifying effect as a result of the natural operation of their propulsive system.

EARTH TECHNOLOGY

When witnessing a space shuttle launch, it is hard to believe that Earth technologies are still very crude. However, all our really powerful systems can be reduced to the 'action–reaction' principle. We rely heavily on systems which amplify pressure to propel our cars, boats, planes and trains through the expansion of gas in semi-sealed or variable-volume containers. At the moment, the compounds used are mainly fossil derivatives or compounds expertly distilled by chemical means. These systems do produce a lot of power but they are incredibly inefficient and tend to waste a very high percentage of the power that is produced in the form of heat and friction. These power sources are also violent and explosive. They are barely controllable unless sophisticated systems are used, such as those found in the modern jet aircraft. Of course, things do go wrong in even the most advanced machines, and when they do the cost in human life is generally high. It is therefore easy to see what advantages there would be in a system which does not rely on such brute force, waste and risk to the user.

While huge electrical generators produce the energy we all desire so greedily in ever-increasing dosages, these same generators are propelled by equally crude systems which employ oil, coal, gas, hydro-electric systems or nuclear steam turbines to drive them. A few systems now produce the odd kilowatt or two by wind power and incongruous 'wind farms' can be seen blotting landscapes which, paradoxically, are usually the prettiest in the land. Despite the quest for so called 'free energy', we are collectively no closer to solving the almost insurmountable problem of balancing energy production with the cost to the health of our planet. Imagine, therefore, the excitement of finding a new limitless, cheap and environmentally clean source of energy.

Scientists and non-scientists alike have searched for this dream and many 'New Age' researchers believe that it is only a question of time before we may all live in the age of Aquarius, free of the shackles which bind us so tightly to our old ways. Many say that they have rediscovered old principles that were once lost, and reappraisals of the work done in this field by Nikola Tesla, John Keely, Dr Moray, Townsend Brown, Wilbert Smith and a host of others are being made by groups all over the world. Claims of 'over unity' electrical devices (which produce more power than they require for their operation) are made regularly. Sometimes construction details are forthcoming, so that experimenters can build their own machines to test themselves.

There are allegedly cases on record of people building machines or devices in accordance with ET instructions, and it is claimed that such objects are far in advance of current Earth technologies. Unfortunately for humankind, it seems the devices nearly always self-destruct, with no remaining traces or even clues as to their construction, so it is difficult to confirm or disprove claims.

Power sources have always held a fascination for human beings, from the majesty of nature to our harnessing of steam, the internal combustion engine and the mighty rockets that power us and our artificial satellites into space. However, few of our achievements are as awesome and terrifying as a thermonuclear blast. Imagine the excitement, therefore, when we get our hands on an alien disc's power source, a system that could possibly transport people and cargo across star systems and galactic space–time, but at the same time have a more parochial use.

FLYING DISCS

In the 1950s, speculation revolved around observation of the flying discs' behaviour in our atmosphere. In flight, they would sometimes change colour from green to red, red to orange, then white, then blue, depending on whether they were stationary, accelerating or travelling steadily. Some researchers thought that the colours represented energy levels somehow expressed in the skin of the disc, others that it might be due to an ionizing effect caused by the actual energy source, which they believed to be a device that could somehow neutralize gravity in much the same way as Jules Verne had proposed.

While it may not be generally known, levitation of objects using current technology is not unprecedented and there are well-documented experiments conducted by Thomas Townsend Brown in the US which demonstrate the Biefield-Brown effect, an electrical field phenomenon which causes devices to levitate. I understand that the project peaked in the mid-1950s, after which Brown faded from view. It was said that very large voltages – in the order of millions of volts – were employed in the experiments (Nikola Tesla also experimented with very large voltages – see page 111) and that much of the work was eventually classified.

John Ernst Worrell Keely was a turn-of-the-century experimenter. However, in his case he appeared not to work with extreme voltages but tended to take advantage of what he called 'sympathetic vibrations'. Work still continues in an attempt to understand and hopefully duplicate some of his experiments, despite the fact that Tesla denounced him as a hoaxer.

There is some evidence to suggest that NASA have experimented (and probably still are experimenting) with 'anti-gravity' using very esoteric acoustic systems, and papers have been published to verify the fact. The creation of micro non-gravity systems on Earth is of obvious benefit to NASA and their ongoing research.

The Nazis supposedly explored the possibility of using flying discs. They were inspired by what they gleaned from the strange occult Thule, Black Sun or Vril Societies. There is strong documentary evidence to suggest that by the end of World War II the Germans did possess radically different craft from anything else the world had seen. Claimed to be powered by 'gravitational engines', these disc-shaped craft bore a remarkable resemblance to the 'modern' discs that are seen in our skies, and photographic evidence does show flying prototypes. Apart from using an extremely esoteric gravitational power source, the Nazis were said also to have used a number of large BMW engines driving huge fans to power other discs, and remarkable specifications and performance claims were made which were unheard of at the time. The eventual and, some would say, inevitable addition of the odd machine gun underslung from the curved belly of the disc stamped the Nazi authority and trademark on the device. Thankfully, there is no case on record of any of these discs being used in war, so we must consider that they were experimental prototypes.

The Thule (pronounced 'toolee') Society appears to have been the driving force behind the neo-Nazi Party. Their interest lingered in the tales of Nordic ancestry, myths and legends which Alfred Rosenberg related in his 1930 book *The Myth of the Twentieth Century*. Because of his connections with the Thule Society, he became 'Deputy to the Führer of the National Socialist Party for the Entire Spiritual and Ideological Training of the Party'. The Thule Society believed that entrances to the 'hollow Earth' (see Appendix, page 246) lay at the North and South Polar regions. Some years after the war, an Admiral Byrd claimed that he had flown over a vast area of what appeared to be sub-tropical land just a few hundred kilometres from the South Pole. While the claim seems to be somewhat fanciful, there are records to show that near the end of the war several Nazi submarines transported considerable cargoes towards Antarctica.

A strange postscript to all this concerns an alleged 'invasion' of Antarctica by an American task force in the late 1940s. The reason for the invasion was allegedly to wipe out a secret Nazi base from which the discs would be manufactured and operations could be launched against their enemies. It is alleged that the American commander lost many aircraft in the operation and had to retire completely within eight weeks, despite being given a lot longer to finish the job. The problem with all of this conjecture, theory and evidence is that if events actually happened as described it would mean that the US got its hands on gravitational drive technology at least two years before the alleged Roswell crash of an alien disc. It could also mean that later claims of the possession of advanced technology by the US (which we shall soon explore) may have been compromised. Actual evidence of the existence of these advanced machines could lie hidden in some vault or cave, or even be broken down into packing crates in a dusty and forgotten warehouse, just waiting to be rediscovered.

Some students of this rather bizarre episode in history maintain that the reason that such radical new devices could be made was because of contact with the Nazi occult advisers, via mediums channelling from the home planet of the Sumerian civilization in the Aldebaran star system (about fifty-seven light years from Earth in the constellation of Taurus). The information imparted by the Sumerians (or aliens?) would have enabled the Nazis to build invincible weapons and fulfil the old Eastern prophecy of a new Babylon. Were these strange flying discs Hitler's famed secret weapons, or was it all occult nonsense?

The Nazis may not actually have got that far with their building plans, which included ideas for enormous machines housing smaller discs that could be deployed at will and protective underground bases from which the discs could be launched. The Allies certainly had some knowledge of the Nazi research and great efforts were made to try to discover the status of the various projects, one of the results of which was to flatten the Peenemunde facility on the Baltic which was used for 'V' weapons research and development. Fortunately, the war was in its final stages, and under Operation Paperclip the US repatriated a great deal of German research back to its own scientists. The most visible results of this culminated in the extraordinarily successful missions to the Moon, which probably would not have occurred within the timeframe without the use of German research and the particular guidance and brilliance of

Werner Von Braun.

P O S T - W A R   D E V E L O P M E N T S

The 1950s were very formative years for ufology in the US. The war in both Europe and the Pacific had ended and there was much reconstruction to be carried out. A great release of imaginative and creative energy burst across the free world, buoyed up by the individual's thankfulness and surprise at actually surviving such a terrible time and emerging more or less intact. Numerous very poor black-and-white science fiction 'B' movies were made, based on what people thought they saw in the post-war skies. Many people still remembered the panic that the *War of the Worlds* radio broadcast had caused in the 1930s (see page 181), so this type of movie entertainment was popular. World War II pilot encounters with the unexplained 'foo-fighters' had further increased the public's awareness of aerial phenomena.

Towards the mid-1950s a man named George Adamski claimed that he had met an alien from Venus in the Californian desert. Signed affidavits from witnesses affirmed that Adamski had appeared (solo and at some distance) to meet someone in the desert. Wonderfully clear photos of a Venusian 'scout ship' were published. The ship was almost identical to some of the Nazi experimental craft – is it possible that Adamski saw one of the Vril or Thule discs with 'gravitational engines' which had been brought back to the US after the war and were being tested in the desert? Or maybe he saw some captured Nazi plans or data? Adamski had also taken other photographs through his telescope of 'mother ships' with 'scout ships' in attendance. Had the Americans continued the aborted Nazi work and did they actually build the huge ship that the Nazis had planned so many years earlier? Had the flight crew been instructed to tell witnesses to their test flights that they came from Venus, so as to divert attention from their real purpose? Adamski was later to produce cine film of airborne 'scout ships'.

Dates may be significant here. The war ended in 1945 and the Allies picked over the remains of Nazi Germany, eager to get their hands on German technology. The US acquired Werner Von Braun and masses of weapons data relating to rocketry, guidance systems and other technical matters that could easily be translated into their own ambitious post-war aerospace industries to keep America commercially in front of the world. This data had been amassed over several valuable years of work by scientists at the leading edge of their technologies in both pre-Nazi and Nazi Germany. The Americans also took the less important but nevertheless interesting research relating to the occult disc experiments. The US then investigated and tested the Nazi discs and realized their great potential when they found that they actually worked. However, there was a serious problem – what if someone witnessed a test flight? The idea of disinformation was born.

In 1952 the scheme was put into operation in the Californian desert, and a man named Adamski – and through him, a whole nation – was duped. It had taken only seven years to translate a Nazi machine which would undoubtedly have been used for oppression into a propaganda coup by means of a peaceful rendezvous in a Californian desert. The question that must be asked is: was Adamski used as a dupe (or agent) by the US government to spread wild stories about the discs that nobody would believe, in order to divert any serious interest from the American public in a covert programme to produce the Nazi machine? It is, of course, all conjecture and there is no hard evidence to support the theory, but it is interesting to explore the possibility that something like this may actually have happened. Until our political masters let us in on the secret we shall not know the truth.

Books were written and Adamski was in demand for lecture tours. He claimed that he was taken to other planets, met Jesus and was imparted much information on the philosophy of the Venusians and their way of life. Messages were given to humankind to mend its ways and the whole thing became something of a circus.

Adamski was called 'Dr' or 'Professor' by his adherents, and a 'snack bar owner on the side of Mount Palomar' by his detractors. Little headway seems to have been made with regard to the scout ships' propulsion system, except to conclude that it was some kind of anti-gravity generator set-up – it is a pity that Adamski was not more technically minded. A medium called 'The Tibetan' did state that a better understanding of what made the saucers 'tick' would be possible when we had a better understanding of the relationships of colours to frequencies. (Interestingly, the early American 'SAGE' radar operated on a frequency of 425–450Mhz, which was claimed by some ufonauts to have a deleterious effect on their on-board systems – it knocked them out of the sky! As a strange aside, the SAGE (early defence radars) frequencies were allegedly used at Montauk to carry out bizarre high-energy experiments into mood and mind control during the 1970s and up to summer 1983 – see Appendix, page 247). Quite what The Tibetan's statement meant is open to liberal interpretation.

While the Adamski episode probably did little to shed any real light on the subject of UFOs, what it did do was focus the American public's minds on a phenomenon which would henceforth occupy much more of their thoughts and energies in trying to solve the mystery. This was in direct opposition to the possible intentions of the US government – to divert interest *away* from these strange airborne objects.

UNDERSTANDING THE PHYSICS

Some scientists with the courage to explore less conventional physics have postulated that all matter is surrounded and permeated by a kind of dense particle or waveform 'soup' (depending on your view). This 'soup' has the interchangeable names of neutrino field, graviton field or tachyon field. Physicists have recently discovered that the speed of light is not as constant as we would imagine and that variations do occur, particularly when planetary bodies distort space. We should not be surprised if zero-mass photons are affected equally by the relativity rules which affect everything else. This is measurable when planets are in conjunction with each other. Some scientists theorize that gravity is not an 'attractive' force but rather a 'push' from outside. This is postulated because the dense 'soup' will tend to force free particles or energy into homogeneous stable structures or, if you like, consolidated matter, which eventually forms the stars and their attendant planets. It is also postulated that the 'soup' possesses immense energy and it is this power that our ufonauts tap to propel them vast distances across space and time, transcending the speed of light. Could mass be merely a 'knot' in the tachyon fabric?

Some propulsion theorists suggest that the only way the discs could perform violent manoeuvres such as right-angled turns at over 1,600kmph (1,000mph) would be to have a virtually massless vehicle. In 1985 Kenneth Behrendt postulated such a theory (see Bibliography). He called it the 'AMF' or 'Anti Mass Field' theory. The theory presumes that all of the sub-atomic particles in the atoms that comprise an object emit or radiate a kind of non-electromagnetic radiation that expands away from the object at the speed of light. He postulates that it is the refraction or bending of that radiation as it interacts with the radiation from other bodies which produces the effect which we know as gravity.

When an object moves, its motion forces the invisible radiation it emits to bend and this bending causes a force we know as inertia, which acts against the acceleration or deceleration of objects. Behrendt states that contemporary physicists may refer to the unseen radiation emitted from the sub-atomic particles of an object as the object's 'gravity field'. However, he prefers to call it the object's 'mass field'. He postulates that if the mass field could somehow be attenuated or partially neutralized, the subject object would lose some of its gravitational and inertial properties and would therefore effectively have a lower mass. It follows that if all the mass field of an object is extinguished, that object would become totally massless. Behrendt theorizes that the ufonauts have perfected anti-mass generators to power their craft and it would seem that his hypothesis has much to commend it, although there are several areas which cause concern, such as what would be the immediate effect, short and long term, on living biological entities, and how would the power levels required for interplanetary flight be generated by a small on-board power source?

BOB LAZAR

On 12 September 1989 George Knapp, host/producer of the KLAS-TV programme *On the Record*, interviewed a young man named Robert Lazar. Bob Lazar claimed that he had worked as a physicist at a very secret and secure government facility in Nevada known as 'S4' (part of the Area 51 complex), in a 'back engineering' project (taking apart an object to find out how it works) on a flying disc. He was convinced that the technology was not of any Earth manufacture, particularly as the disc utilized physics unknown to Earth scientists.

Bob Lazar's claims were, and still are, quite astounding. He is a somewhat colourful character in the normally staid world of physics, having built jet-powered motorcycles and cars before getting involved in this extremely sensitive government project. His background is equally colourful: he was convicted on 20 August 1990 of pandering (living off the earnings of a prostitute) and was sentenced to a three-year suspended sentence and 150 hours of community service. In fact, he was not caught 'bang to rights': his conviction and subsequent punishment were administered basically for installing video equipment and helping with computer software and accounting for a female friend who owned an illegal brothel in Las Vegas. In that respect, Lazar could not be considered a hardened malefactor, especially as he had no previous convictions. However, he had admitted previously that, together with his first wife Carol, he had owned a legal brothel called the Honeysuckle Ranch in northern Nevada. His need to earn money no doubt had something to do with the fact that he had recently lost his job with Naval Intelligence. However, this unfortunate brush with the law did not enhance his credibility, particularly as he now claimed that he had worked on probably the most secret project the world has ever known.

Bob Lazar claimed he had worked at the Nevada test site near the Groom Lake area for no more than six or seven days between December 1988 and April 1989. During that time he had been given around 120 briefing papers to read, presumably to bring him up to speed, in which he discovered that the US had acquired several alien craft whose occupants had originated from Zeta Reticuli 4 (the fourth planet out from Zeta Reticuli – see Appendix, page 244). The technology that had enabled them to make their journey used a power plant capable of generating enormous gravitational fields, in a process where matter was annihilated in a matter/anti-matter reactor to produce prodigious amounts of electrical energy in a thermal electric couple, which Lazar said was nearly 100 per cent efficient.

The reactor
The reactor utilized small pieces of a very heavy but stable element which Lazar called element 115. This was placed in the on-board reactor and bombarded with protons. The material then transmuted naturally to element 116, which almost immediately decayed and released anti-matter. The anti-matter was then released into a tuned tube, which isolated the very reactive material into a gaseous target at the end of it. The result of all this is to produce an intense gravitational field (Lazar said the nearest thing it could be likened to is a 'black hole') which is out of phase with the disc's environment.

Lazar contends that he now understands that gravity is a wave and not a particle (the graviton), as previously thought. (Physicists are now tentatively coming around to this way of thinking, with 'wavicles'.) Element 115, which does not exist on Earth – nor can it be synthesized here – produces an excess of the gravity wave (note the

similarity to Behrendt's 1985 AMF theory) due to its nuclear configuration. This excess energy from the wave is tapped, and amplified to
produce the enormous field strengths to power the disc. Lazar
believes that there is about 227kg (500lb) of element 115 available for
very secret research (in the US only?); it is very heavy, coloured
orange (yellow-orange like uranium ore?) and melts at 1,740°C.

One of the remarkable aspects of the reactor described by Lazar is
that there is no switch to throw or buttons to push to get it working.
Once the 223g (7½oz) triangular-machined piece of element 115 is
placed in the reactor, power is produced immediately. (A problem
here is that very little anti-matter has ever been made in laboratories
and even nature makes only very small amounts. The continuous total
annihilation process should produce heavy gamma emissions, which
would be lethal to the occupants of the disc and anyone else in the
immediate vicinity of a working reactor.)

Flying the disc
One of the effects of the disc in operation is to distort space. This
effect can cause the disc to become invisible if it is viewed from a particular angle and the on-board gravity amplifiers are in a particular
configuration. Lazar claimed that he actually witnessed a short flight
of the disc he had been working on – just up and down, but he maintained that it was incredibly impressive to see it rise almost noise-
lessly, hover and then sit down again. He called it the 'sport model'
because of its sleek appearance. Puzzlingly, it is almost identical in
appearance to 'Billy' Meier's 'beamship' (see page 18).

Lazar explained publicly how these discs negotiated the vast distances without violating what we know of relativistic physics. The
gravity amplifiers of the disc's power plant are focused on the point
to which it is to go. The gravity amplifiers are then switched on, bending or warping space and thereby affecting time as well. The amplifiers are then switched off and the disc is carried along with
space–time as it retracts. In this way, no linear travel is made and to
all intents and purposes time is largely suspended, which means that
vast distances may be travelled. However, Lazar did not elaborate on
the navigation aspects of traversing stupendous distances in the
search for individual planets within a star system. Nor, from his
description of the 'sport model' interior, is there any indication of
navigation, sophisticated or otherwise. However, he said that much of
the interior of the disc had been stripped out, leaving very small

chairs (too small for adult humans), the reactor and some other hardware. He did not explore the upper compartment of the disc as his brief concerned only the power plant. On a planet, such as Earth, the disc cannot use the gravity amplifiers in the same way. Instead, it is balanced on Earth's gravity by an out-of-phase configuration which is analogous to a 'cork on water'. Lazar maintained that in this configuration the disc is subject to the vagaries of planetary weather.

Technical questions
In an interview with KVEG Radio on 28 December 1989, Lazar elaborated on some of the more technical aspects and answered questions posed by phone-in callers. He was asked if there was any thermal radiation when the gravity generators were in use. He said that while he had never been in the location of the gravity generators while they were running and therefore did not know whether thermal radiation was in evidence, he had noticed that the reactor did not heat up at all. He found that astounding, because it contradicts the First Law of Thermodynamics. Lazar said that because of the huge power generated there is no direct physical coupling between the gravity amplifiers and the reactor, and the device seems to operate like a Tesla coil with components being tuned to one another. Interestingly for physics students, he said that the actual gravity wave straight out of the reactor is of far too low an amplitude and is only of use once it has been amplified by the gravity amplifiers. The reactor itself is a small plate about 45cm (18in) square, with a half sphere placed on top of the plate to tap off the gravity wave. In one demonstration Lazar witnessed the repulsion field from a working reactor, which was much the same as the two opposing poles of a magnet. He also witnessed an intense gravitational field that had been focused and, intriguingly, saw a small black disc appear due to the bending of the light around the area of maximum focus.

In other interviews, Lazar confirmed that the gravity amplifiers always run at 100 per cent and the resultant gravity wave is phase shifted between zero and 180 degrees. Most of the time it is at a null setting. He mentioned that the reactor runs at about 7.46Hz (around the natural frequency of Earth – the natural frequency of gravity perhaps?). The system operates in pulses, not a continuous stream, and Lazar maintains that gravity is just part of the general electromagnetic spectrum. Unfortunately, conventional science has not woken up to it yet. Asked what the actual frequency was, he demurred, but indicated it might be in the microwave range and that the waveguide mechanism in the disc would yield clues. He admitted that he was currently working on the problem on his own account and could not give away more information at the time.

What is going on?

Lazar commented that he considered the level of expertise and investigative scientific procedures at S4 to be very amateurish. A minimum of equipment seemed to be in use and consisted mainly of volt meters and oscilloscopes. Considering the very advanced level of the technology being investigated, he found the *laissez-faire* attitude of the supervisory staff ('try a procedure if you want to') incredible and was amazed that more care and greater enthusiasm was not shown. He maintained that workers were not allowed to communicate with each other and there was a very high proportion of guards to personnel. He was also convinced that there were attempts to control workers by the use of drugs: he claims that he was once made to drink a yellow liquid that smelt of pine and feels that this substance was administered in order to control the minds of personnel at the base.

Lazar speculates that he was brought into the project in order to replace one of three scientists who lost their lives during an underground test in May 1987 when they attempted, somewhat unwisely, to dismantle a working reactor. The resulting detonation is supposed to have vaporized the immediate area and ripped off a blast door which hardly ever breaks. He believes that the US is trying to duplicate the alien reactor using plutonium as the fuel, which unfortunately does not work too well (shades of the 1980 Betty Cash incident? See page 67). He also believes that the US is very interested in the weapons aspect of the research, particularly as an anti-matter bomb would make the most powerful H-bomb look like a damp squib!

The Star Wars programme (Strategic Defense Initiative – SDI) may be funding the 'black projects' at S4 and the search for exotic propulsion will go on. It has been rumoured that the latest generation of stealth bombers actually have a mixed propulsion system – radar-cloaked jets and the alien reactor – which allows the aircraft to perform manoeuvres beyond any known performance envelope. There is a problem with this, however: what if a bomber carrying one of these power plants crashes in a foreign land – surely that would hand the recipient nation technologies beyond their wildest dreams? It has been alleged by some sources that the US continues to experiment with radar *and sight* invisibility for their stealth systems. The 'electromagnetic bottle' design may now be obsolete if radar and sight invisibility can be achieved with Lazar's anti-matter reactor.

While Lazar was hired to work on the disc's propulsion systems, he did read briefing papers on other subjects. Some of the data concerned projects connected with particle weapons and there was even one concerning time travel. He also read accounts that the aliens had manipulated human beings' evolutionary progress some sixty-five times in our history to date. Lazar maintains that he saw a very thick dossier marked 'Religion', which he did not read. However, if the inference is that Earth religions are an alien construct there might be many atheists of all nationalities who would find an endorsement from outsiders rather superfluous.

Lazar claims that his short stay at S4 was not happy and he hated the work environment. He was escorted everywhere and on occasion was shouted at in a manner which would be more in line with Green Beret training than a top secret government facility. That, and the possibility of the use of mind-control drugs which were administered without consent, made him determined to leave S4 and go public. He maintains that he was threatened with being shot and his body dumped in the desert where no one would find him, if he did not behave. Lazar concluded that going public would be his best defence. If subsequently anything did happen to him people would know about S4 and its regime – and that it was the US government who had eliminated him.

Is it true?

Lazar's story does not diminish in its telling or with the passage of time. It is as exciting and compelling now as it was when he first went public in 1989. He has been consistent in the telling and when asked questions he does not know the answer to he has said as much. He has undergone lie detector tests and passed them without difficulty, all of which tends to substantiate his story.

However, there are some, including Dr Jacques Vallée, who believe that Lazar was in some way used and that the S4 scenario indicates pure theatre. The question is, for what possible purpose? There remain innumerable inconsistencies which relate specifically to the scientific (or, as Lazar claims, *non-scientific*) approach to the work at S4, including the clear lack of serious and sophisticated instrumentation for studying the subject in a correct scientific manner. This aspect probably constitutes one of the main objections to Lazar's story and is a repeated stumbling block for serious consideration of his claims. There is also the problem with the known characteristics of anti-matter itself. Very little exists artificially – for example, in particle accelerators – or in nature as a result of colliding particles. Anti-proton or anti-neutron collisions with their counterparts would release very high gamma radiation, which would be lethal for any living organism near enough to the source. There is currently no shielding method known (to our science) which could be considered reasonable, especially for a flying craft.

The non-scientific approach at S4 could mean that the human participants are not actually in charge and are merely operating under instructions. While Lazar is adamant that he did not see any ETs at S4, he did say that on one occasion he saw two lab-coated individuals facing him and looking down, talking to something small with long arms through an observation-type window in a door. Lazar maintains that while the incident may suggest that a grey was being spoken to, he cannot confirm that. He emphasizes that he caught only a glimpse of the scene in passing, so he cannot confirm that there are aliens at S4 working with American technicians and scientists.

The Groom Lake area has certainly been used for secret projects before and these include the SR71 (Blackbird), the U2 spy plane, various stealth projects and now, it is said, the top secret Aurora project. In that respect, S4 would be an ideal place in which to carry out experimental flights of alien craft. Since Bob Lazar first told his story, countless numbers of interested spectators have gathered at the Freedom Ridge and White Sands viewing areas to capture on still and video cameras the erratic flights of craft the US government continues to say do not exist. Unfortunately, the land has now been seized by the government in the interests of national security, probably because of the serious interest shown by commercial media companies. While some of the cavorting lights in the night sky may be RPVs, it is unlikely that the authorities would develop such a tight security net around the area without there being something very special to protect.

When the Lazar story broke in 1989, it exploded like a bombshell in the UFO community. No doubt such a detailed account from an eyewitness, who had not only worked on an alien disc but had also seen that technology demonstrated, would have given the US authorities something to think about. If what Lazar has said is true, they would seem to have two options: tough it out and hope that it will go away, or spill the beans and tell the world what has been going on since 1947, or maybe even before 1947. The first option is the one that will be taken, for it is probably not in the US government's interests to disclose the full account of this aspect of history.

The high probability that the US Congress has no knowledge of these areas of operation will add to the reluctance of those running these extremely exotic projects to go public, even if they could. While there is no actual proof that information from the projects has been disseminated to other world powers, it is very likely to be the case, with the British government included in a share of some of the data. However, the British people will probably be among the last to know due to their government's obsession with secrecy.

ALIEN ARTEFACTS

The general subject of hardware is a very vexed one for the serious investigator, particularly as possession of an actual alien artefact, whether it be a flying disc or even just a piece of such a vehicle, would prove incontrovertibly that aliens do exist. Unfortunately, there are many people who believe that everything that falls from the sky must be some kind of alien artefact; the celebrated Lonnie Zamora case is a possible instance of mis-identification.

The incident occurred on 24 April 1964, which places it squarely in the heyday of the US space programme. The description of a roaring sound and flames jetting from a shiny, egg-shaped vehicle does not suggest alien technology but rather an effort on the part of the US programme to produce a manned VTOL (Vertical Take-off and Landing) machine. The description of small beings piloting the vehicle gives one slight cause to reflect, until you realize that small pilots would be selected if the power-to-weight ratio of the craft was to be kept under control. One would, of course, expect landing traces and scorch marks from rocket or jet exhausts from such a terrestrial machine. Dr Hynek reviewed the case for Project Blue Book and concluded that this case clearly suggested a 'nuts and bolts' craft, which was patently true. However, I suspect that the nuts and bolts originated slightly closer to home than many others thought.

The Kecksburg incident was another very strange affair which was reported to have taken place in the late afternoon of 9 December 1965. A strange fireball was seen to fall into a wooded area near the small Pennsylvanian town of Kecksburg in Westmoreland County. It has been suggested that the US Army, Navy, Air Force and NASA all contributed to the recovery and inspection of the device (which presumably is still in the hands of the US government). Judging from a model of the strange belted acorn shape built by NBC-TV for a programme called *Unsolved Mysteries*, it was certainly large enough to carry normal-sized human beings.

The case is also intriguing in that witnesses to the event were still coming forward some twenty-five years after the event, stating that they had seen the object performing slow manoeuvres over the site of the landing, with controlled descent into the woods. However, it is understood that investigations by those researching this occurrence have turned up several anomalies which remain today, one of which must be the strange writing (or what is presumed to be writing) around the belted base of the object.

The capsule/object – call it what you will – bears a remarkable, if superficial, resemblance to an Atlas re-entry type of vehicle. As such, it could have been an experimental re-entry capsule constructed by either the US or the Russians, who were engaged in a prestige space race at the time. It should be remembered that the first American manned space launch occurred on 5 May 1961, followed by John Glenn's flight on 20 February 1962 and Scott Carpenter's on 24 May 1962. The precursor to all of this was, of course, Juri Gagarin's historic and brave first orbital flight in 1961. It follows that the US military machine and NASA would be interested in recovering their own vehicle, but especially interested in recovering that of a competitor.

It is not unrealistic to assume that *all* of the space junk that falls to Earth is of terrestrial origin, particularly as the amount of garbage orbiting the Earth is increasing almost daily. Following the old adage 'what goes up must come down', one could conclude that the Kecksburg object might have been such an item of man-made origins.

Unfortunately, the truth will probably never be told. The problem surrounding all cases of so-called alien artefacts is that hardly any details are known. The exception is the Bob Lazar case, despite the confusion and doubt over his bona fides. However, there is still no hard evidence at all, merely *claims* that hard evidence has been seen, and we need to go beyond blind faith to progress the subject further. The tantalizing story of 'Billy' Meier and his metal samples which were lost or stolen (see page 18) is another example of a nearly proven case: the subsequent events which tended to discredit Meier may have been put in place to dampen down the whole issue. The Adamski photographs, 'proven' by orthographic projection techniques and the claims of experts that they were real, still does not come up with hard evidence. It is strange that the 'Nordic' type of alien visited upon Adamski now appears to attract the attention of well-established cult groups. It is also strange that some have said that Adamski possessed a special type of US passport which would only be issued to dignitaries or officials, thus allowing him the convenience of uninterrupted travel, anywhere. Perhaps if George Adamski was working for the CIA he did achieve his mission objective – to confuse the enemy and create obfuscation. The obfuscation remains to this day.

MISSION LIFE?

There is no difference between time and any of the three dimensions of Space, except that our consciousness moves along it

The Time Machine
H. G. Wells

The Earth, the accepted nine known planets and our Sun are located on the outer limits of one of the spiral arms of the galaxy known as the Milky Way. This is an enormous (the word hardly seems adequate) cluster of stars, gas and stellar dust a few hundred light years thick and at least 10,000 light years across. It is around fifty times as wide as it is thick. Approximately one billion stars make up the Milky Way, with the bulge of Sagittarius in the central yoke and our Sun near the edge. The system rotates slowly around the central yoke, thus holding the entire galaxy in a gravitational embrace.

As everyone knows, the Earth is predominantly a water planet and is the only example of its type in our solar system. Despite its beauty and diversity, it is very small and insignificant when compared to the Sun and many other of the planets. For example, if you had a circular field some 4km (2½ miles) in diameter and you placed a ball around 60cm (2ft) in diameter in the middle of the field to represent the Sun, you would place a mustard seed around 25m (27yds) from the centre to represent the planet Mercury. You would next place two green peas 43m (47yds) and 66m (72yds) from the centre to represent the planets Venus and Earth. A large pinhead at around 100m (109yds) would represent Mars, and a tangerine orange about 0.4km (¼ mile) away would represent Jupiter. A small spherical lemon almost 0.8km (½ mile) away would represent Saturn, a large cherry

about 1.2km (¾ mile) away Uranus and a plum at the edge of the field
Neptune. Pluto would be about 0.5km (⅓ mile) further out from
Neptune and would be approximately the size of a small pea. The
solar system is constantly subjected to bombardment by meteorites,
meteors, comets and other space debris.

Even though the galaxy is thought to contain over one billion
stars, it is merely one of many similar galaxies scattered throughout
the Universe. In fact, there may be as many as a hundred million separate galaxies in the Universe and it appears that (in accordance with
the 'Big Bang' theory) they are receding from us.

FINITE OR INFINITE?

Scientific debate still rages as to whether or not our Universe is a
'closed' (finite) or 'open' (infinite) system. A problem occurs
when trying to calculate its density. Einstein produced equations that
can be used to calculate the curvature of the Universe, based on the
principle that mass will cause the space around an object to distort. It
follows that if the Universe could be weighed (by a mass inventory),
the curvature due to gravitational field distortions could be predicted
and would therefore indicate whether or not the Universe was infinite.
Fortunately, on a large scale the Universe seems to be isotropic (it
looks broadly the same in any direction) and there appear to be as
many observable galaxies and spectra in the radiation bands
detectable by us, which makes predictions of an open Universe more
probable than improbable. (Details are given in *The Hidden Universe*
by Michael Disney – see Bibliography.)

Einstein predicted mathematically that there is a critical density of
the Universe and that if the actual mass density of space is greater
than the critical value, then the Universe is 'closed'. It follows that if
the density of space is less than the critical value, then the Universe
is curved, but 'open' and therefore infinite. Based on evidence to date,
mass inventories of the Universe and calculations from those studies
prove that *observed* densities are only about one per cent of the critical density to close the Universe. The problem arises because of the
word *observed*. Cosmologists are concerned that there may be missing masses many times over what they can see and record. Apart from
the notoriously difficult task of weighing galaxies, a great deal of
additional mass may exist as neutrinos (although these are considered
virtually massless, their theoretical proliferation throughout the

Universe making up for their lack of mass), radiation, anti-matter, intergalactic gas and gravitational waves. The quest for this so-called 'black matter' has also given rise to the idea that energy itself has mass. If you accept that 'energy is but frozen matter' (Sir James Jeans), the idea has possibilities.

If any of these aspects is left out of the equation, it will tend to distort badly any final results or conclusions. The difficulty is, how do you include them in the first place? The 'Big Bang' theory sits uncomfortably with some scientists because of the difficulty with the 'Singularity', the fact that no one has satisfactorily predicted pre-'Big Bang' and the problem with the missing mass (observations suggest that there is at least a hundred times more mass in the Universe than can be accounted for by the visible galaxies). Add to this the mathematical proof that galaxies cannot condense out of an expanding Universe (Lifschitz, 1944) and it becomes clear that in the face of no particular all-embracing solution to the problem, the creationalists probably have as much right to their 'solution' as do their scientific opponents.

The origin of sentient life must by definition be linked to the life of our Universe and the system which supports it. The Earth sustains us, but we, along with everything else in our Universe, owe our origins to the stars. The initial creative process may have been a 'Big Bang' or even an energetic 'Push', followed by a comparatively 'Steady State' system so that the process of creation can be sustained. It is inconceivable that one climactic and incredibly violent act would provide such impetus to a continuum which to all intents and purposes would last for what we call eternity.

CREATION

When Charles Darwin shocked Victorian Britain by stating that humankind was an evolutionary product from ape stock and could not therefore be a divine being created directly by the Hand of God, he stirred up a hornet's nest. Nevertheless, Darwin's Theory of Evolution and Origin of Species persists today in the absence of any other solution. Modern physics has done much to draw back the veil of understanding over our immediate surroundings and our Universe, yet even great innovative and intuitive minds like Dr Stephen Hawking must wonder at the way our Universe seems to act as if it were a living organism, with all the hard-won answers giving rise to even more questions, which in turn provide tantalizing glimpses around the next, apparently endless series of corners.

Perhaps the real answer lies somewhere in between the 'Steady State' theory so eruditely advanced by Fred Hoyle, Hermann Bondi and Tommy Gold in the 1940s, and the current theory of the 'Big Bang' Universe. If our visitors get to us by negotiating space–time using the system of gravitational drive that Bob Lazar describes (see page 77), time or linear distance has no meaning. This implies, at least in principle, that our visitors could travel from adjacent Universes to find sentient life, even if we remain unique in our own Universe. (This presupposes that our visitors do not coexist with us, hidden in an unknown dimension nearer to home; presumably, if *they* do there would be no need of gravitational drive systems for transportation.) However, the method used to chart over unimaginably vast distances is almost terminally incomprehensible to us – even if *they* somehow knew we were here. If the visitors come to us from other dimensions, *they* might originate in Universes far beyond our detection. To explore that possibility we need to carry out a thought experiment. We now enter exciting territory.

W H A T   I F ?

I magine, if you can, an incredibly vast system at rest, with a nucleus of unimaginably dense matter surrounded by countless free particles of immense electrical charge. The scale would be massive, with a nucleus thousands of light years across surrounded by a swimming, swirling, dancing mass of primordial particles, held to an amorphous boundary limited only by the scale of the particle count and maybe hundreds of thousands of light years across in a shifting, shimmering shape (fig 1). The whole system exists in an uneasy equilibrium as a result of the pressure of quantum gravity, quantum particle exchanges and other systems which constantly compete for space (the word 'space' is used outside its normal context here as it is hard to think of an alternative). The system is in constant flux, but the tremendous inertia causes the process to be extremely slow in operation. At any time, stable boundaries and nuclei may become pressurized (fig 2).

While individual cells normally remain discrete, distortion occurs because cells are jostled, squashed and pushed into unstable shapes by the action of other cells (fig 3). Gravitational imbalance is the result and the nucleus becomes unstable. Eventually, the system cannot maintain equilibrium and strange one-time quantum effects take

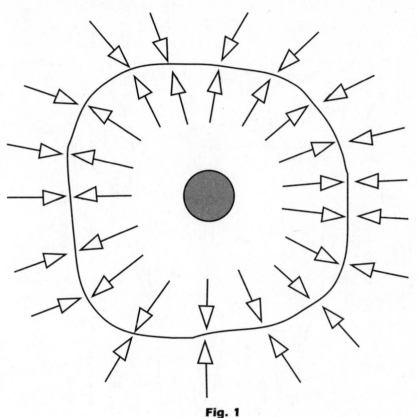

Fig. 1
Pre-'Big Bang' – Stasis (slow time – nil or
imperceptible progression)

place which cause the nucleus to burst. (Due to the random shape of the nucleus at the exact moment of its terminal instability, and the non-homogeneous nature of its boundaries, the probability is that the initial spark of ignition would not be central to the core of the nucleus because of the locality of the pressure wave emanating from the initiating cell or cells, coupled with the actual boundary shape of the deformed cell at the moment of ignition – therefore a burst would not be uniform in every direction). The unimaginably gigantic explosive force of the nucleus throws matter and radiation out into the surrounding unoccupied space in random directions. Vast expanding strands of energy, matter, one-time primordial particles and shock waves expand from the collapsing nucleus at the speed of light, as it

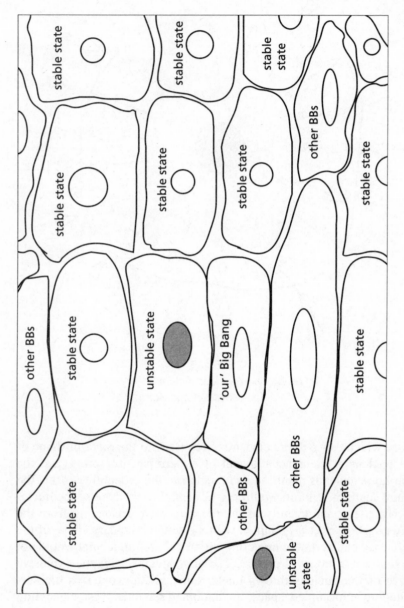

Fig 2 The 'Big Picture'

attempts to re-size itself into equilibrium.

Imagine the scenario if you can. No sound, no light – everything dark except the weak glow from the stasis of the low-energy nucleus. Slow pushes from an adjacent cell – almost imperceptible movements. More pushes from an adjacent cell. The nucleus of our cell suddenly becomes alive, as though awakened from a deep slumber. Radiation emissions increase; sub-atomic exchanges increase. Suddenly there is immense photon emission – visible light where there has never been any before. The nucleus becomes increasingly unstable and starts to move and shift uncomfortably within its boundaries, like a living wild thing trying to escape from a monstrous cage. More pushes. More instability. More radiation and visible light. The light is very bright now, with enormous photon emissions, but there seems no actual source point. It moves, it darts like an unknown lightning trying to discharge its energy in an increasingly energetic environment. Extremely strange one-time anti-particles are created. As fast as a new particle is created, it is transmuted into another new particle or anti-particle. Unsustainable pairs are formed and strong nuclear forces collapse under tremendous quantum gravitational fluxes. The whole process is repeated relentlessly. Relativistic anomalies take place with very young transient wavicles and strings, composed of both gross matter and radiation – conventional physical laws do not yet exist, only probabilities. The temperature of the nucleus suddenly soars to incredible magnitudes. The core starts to emit massive energy in the form of copious emissions, X-rays, gamma rays and the entire spectrum of radiated energy. Boiled-off photons surge from the seething, writhing structure as it tries unsuccessfully to find equilibrium.

The process cannot be sustained for much longer. A spark is generated randomly within the nucleus, caused by a chance concentration of quantum exchanges. This trigger event causes the remaining energetic states to pass the threshold of terminal instability. The nucleus begins to expand exponentially, the location of the igniting spark determining the orientation of the explosion. As the nucleus explodes it produces a display of shimmering, bursting light. It ripples majestically with all the spectrographic colours, and all possible hues of those colours. Whites, golds, blues, yellows, mauves, reds, greens and violets flash and jet from the expended, shrinking nucleus in a boiling, seething photon sea of highly charged elementary particles. All is light where no light existed. Blinding white streams of newly formed gross matter, light years across, burst and escape from

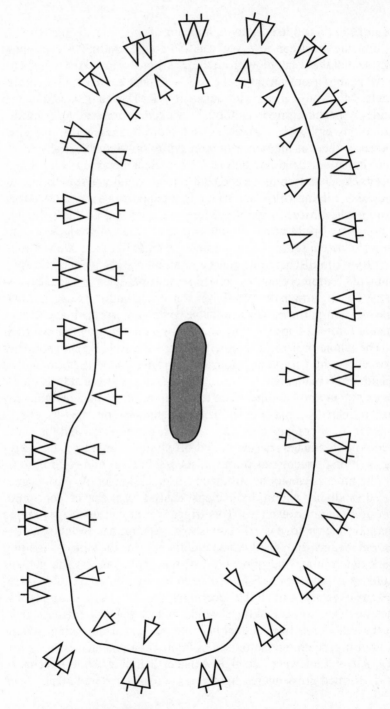

Fig 3 Immediately prior to 'Big Bang' (Time jump – time starts progression)

the uncertainty of the nucleus to seek stability. Some streams will go far, others will wither, according to their energy states. Some streams will be deprived of gross matter and will become Universes comprised of quantum gravitational 'wave memory' radiation and free energy. Some expanding strands carry more matter than others and each becomes a Universe, which is limited in volume only by the mass it has arbitrarily received. Each Universe becomes longer than it is wide because of the acceleration of matter from the expanding core. Sometimes large accelerant strands of matter (immature or runt Universes) will become dynamically intertwined and overlap other accelerant strands for some of their span, thereby allowing the possibility of transdimensional windowing at a later point in space–time.

Sometimes immature weak strands may form from the initial quantum furnace. These puny, doomed strands may overlap other strands, resulting in unknown and extremely strange, anomalistic intra-dimensional behaviour – too strange even to predict (fig 4). Perhaps these Universes exist as a complete opposite of our own, with anti-positrons and anti-neutrons as particle exchanges constantly take place in a struggle to exist. Such a Universe would be a suffocating, hot and irradiated place where total darkness replaces light and nothing we know of in our Universe could survive. Perhaps all Universes formed from this violent birth are 'mass knots' in a thin gruel of anti-photon 'soup'? If these strange and unpredictable places exist we could never go there, even if we had the means, because we would be annihilated by anti-particles.

Expansion of the Universes continues. But, an expansion only occurs if the given mass is large enough to sustain the process. If it is not, the young Universe will tend to shrink back into the remaining primordial vortex, only to be recycled and poured into a more viable Universe which is still expanding (fig 4). Thus, to an observer in an expanding Universe (ours, for instance) it would appear that the momentum of the initial process continues, which of course it does (the 'Red Shift'). However, an observer in a shrinking Universe would have an entirely different perspective – and probably an extremely pessimistic one! The notion that large, expanding Universes feed from smaller, immature Universes also neatly gets around the Lifschitz problem, which proved mathematically that galaxies could not possibly condense out of an expanding Universe.

Anomalistic space–time behaviour would only be apparent to us if we were situated at or near one of these strange events, which could

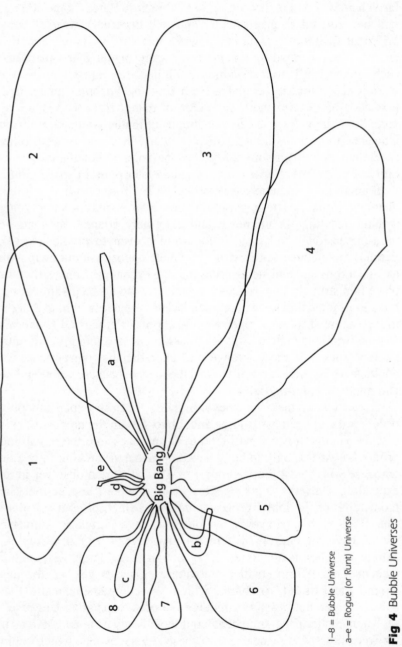

1–8 = Bubble Universe

a–e = Rogue (or Runt) Universe

Fig 4 Bubble Universes

mean being near the edge of our Universe. This is not very probable as it is reasonable to assume from our thought experiment that the matter making up the galaxies would tend to be projected rather more centrally than peripherally. Intertwining behaviour and its effects would only be felt once the expansion processes had slowed markedly or stopped, and would most likely be limited to strange phasing from local distortion of the continuum (an observer's perception of his immediate Universe).

The Universes produced by the 'Big Bang' cannot develop forever. Eventually they must run out of matter to sustain momentum. Whether that occurs before the dimensional boundary of the Universe (interstices between Universes) is reached, or whether expansion is stopped because no more matter is available, may cause one of two quite different effects. In the first case, the Universe might tend to bend around the curve of the boundary of another Universe. In this way it becomes possible for mature Universes to encroach on the dimensional space–time of other mature Universes, with unpredictable effects – perhaps allowing our visitors through to our Universe in such a window of opportunity, if such windows are, to all intents and purposes, permanent features.

However, if the system just runs out of momentum, the Universe in question will simply cease to expand. Such a Universe will not contract but will remain expanded to its maximum size and may eventually become ossified, an incomprehensible fossil frozen in its own space–time matrix once all the sub-atomic energy is depleted and dissipated as radiation. Such a scenario would be grim indeed: an ever more dimly lit Universe with stars going out and galaxies slowly lowering in their light output, as though a giant hand were operating a monstrous dimmer control switch. Such a Universe would freeze all life to extinction long before it became extinct itself. The immature Universes would be the first to go. They would shrink in volume as long as matter is evaporated by proton decay over vast timeframes, something like 10^{300} years, eventually falling back into the weakened cell continuum and existing only as a pale echo of their short life, as the whole system eventually runs down to stasis and extinction.

The final picture of the cell could be similar to a monstrous flattened husk: cold, inactive, spent Universes with matter compressed into small clusters and existing only as dimensional anomalies in their own narrow continua, taking up minimal space until gravitational pressures squeeze them into other boundaries, to be absorbed and recycled via the interstices of adjacent boundary systems in the 'Big Picture' (fig 2). Residual sub-atomic activity may occur but will not be sustainable. The whole scenario would be one of constant creation and recycling, with Universes undergoing peculiar, currently unknown quantum gravitational stages, followed by a period of stasis and eventually extinction, the incredible cycle completed. Using our thought experiment, we can speculate that there may be other 'Big Bangs' occurring in any number of other cells at random points in the creation process, and that each one of myriad cells with its own boundary system may be also enclosed in a greater cell, which in itself is a single entity... and so on. Perhaps creation is more like a Mandelbrot system than we could ever imagine?

Our thought experiment threatens neither God nor the 'Big Bang' theory, and it has no relevance to the 'Bubble Universe' theory put forward by Boltzmann in his thermodynamic model towards the end of the nineteenth century. The 'Steady State' theory of the origin of the Universe has not been fashionable for many years, but our thought experiment takes both the 'Big Bang' and the 'Steady State' theories and brings them closer together. It may also provide some comfort to those more rational thinkers among us who cannot believe, as an act of faith, in eternity – how can we, when we are completely surrounded by finite reality? It may well be that there never was a 'Big Bang' but more of a 'Big Push', where a relatively cool quantum foam burst and high temperatures evolved from quantum gravity heating – who will ever really know? Perhaps *the* event went from 'simple' to 'complicated', from a one-dimensional algorithm containing all the code necessary to produce complex equations through a chance process? Perhaps there should be two new physical rules added to our repertoire of understanding: (a) matter *must* aggregate, and (b) matter *can* aggregate discretely. Maybe God does play dice after all?

Our concept of eternity probably has more to do with our lack of understanding than an implicit belief in what we think we know. Maybe it is a product of our acute and painful knowledge of our own mortality, and the blind hope that something of us survives even if our physical bodies and our many works, after our short span of consciousness, do not. Our fear of death has always tended to distort our perspective, because it means giving up life and the gift of consciousness. Unfortunately, in our world giving up after such a short timespan means evolution for *Homo sapiens* and death for the individual. This is the price of *our* reality. Perhaps the visitors are also bound. In the creational model of our thought experiment there can be no progress by evolutionary means, because the system is already perfect and tuned to its sole purpose – creation. It is therefore a continuous recycling process. The visitors may already know more of the technology of the Creator and our Universe than we do.

TIME

The human perception of time is probably limited by our in-built ability to see time only as a linear progression. The only other natural method available to us for perceiving time in any other way is when we dream. When we are asleep, time does not appear to move at the same rate as when we are awake; it becomes compressed, and it is possible to live out a fantasy which we believe to have taken half an hour or more, when in reality it has been dreamt in a very short timeframe in a burst of brain activity.

We see time as creation, birth and death and therefore convince ourselves that that is how it is. But there are other manifestations of time which generally escape us, like radioactive decay, the levelling of mountains, continental drift... all these occur outside our own meagre timeframe references. On the other hand, there are also timeframe references which are much shorter than ours. For instance, the life of an adult fly is no more than a few short days, and in the case of the mayfly just one day from a hatchling to an adult flying insect. However, the physical laws that bind us in our Universe act on all these subjects in the same way: the arrow of time goes forward, so that the present and past may exist and order may be maintained. Without that order, nothing would exist as we know it.

It is unpopular to talk of time without reference to space–time. In fact, (with the benefit of Einsteinian relativist physics) all physicists agree that it is *incorrect* to talk about time other than in space–time terms. If one argues that radioactive decay performs the classical time-reference feature of progression, it is surely not realistic to state that time did not exist before the 'Big Bang', for there must have been some classical progressions for even primordial particles or short/long strings to develop. This implies creation, birth and eventual death in an immutable congress with the arrow of time. While we have speculated on the state of things pre-'Big Bang', nothing is really known about the state of matter before that event and much is based on educated guesses and conjecture. Our Universe may be much sim-

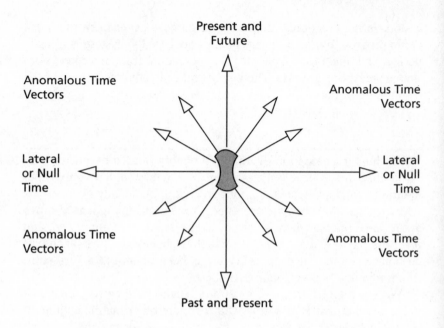

Fig 5 Two-dimensional Schematic of Archaic Time

pler than we think and our visitors no doubt understand more about its nature than we can dream of. Their understanding may therefore provide *them* with the means of intra-dimensional travel and the manipulation of time.

One of the golden tenets of modern physics is that nothing can travel faster than light (although the theoretical particles called tachyons are considered to do so). However, there is nothing to indicate in that simple, direct statement why light itself is not affected by the relativity rules which govern the behaviour of everything else. Perhaps the visitors also understand this problem. Fig 5 describes what may be the nature of our particular time reference. The system works on the precept that the law of progression, if broken, allows the experience of anomalous time references which will not perform classical progression. It must be said that the current understanding of the nature of time does not allow us to explore that prospect and it must remain unattainable for us, at least for the present, even though our visitors may have unlocked time for their own purposes.

The extremely anomalistic nature of the aliens and of the strangeness *they* bring to us may eventually help us to understand the *real* nature of the Universe in which we play a small part. Their eventual overt presence will cause a drastic rethink of our theologies and our accepted beliefs about our own origins, and even about our purpose on Earth. If we can survive the shock, the human race might experience a vast leap in knowledge. If we cannot come to terms with *them* and a probable complete revision of our belief systems, we may not survive the encounter. If *they* turn out to be transdimensional, *they* will already know of many things at which we can only guess. However, whether *they* use gravitational drives or are intra-dimensional travellers, is really only an absorbing adjunct to *their* interest in us and our only home in the known Universe. It is therefore time to explore the possible reasons for this interest in greater detail. There may, of course, be little reason why we should remain optimistic.

THE WATER PLANET

Four-fifths of the beautiful Earth is water. Because of this, it is a unique planet in our solar system. While Mars seems to have had large quantities of water in her past, no other planet in our solar system has an ecosystem capable of sustaining the abundance and diversity of life that exists on Earth. It was once thought that the Earth's sister planet Venus may have had a similar climate and ecosystem, but it clearly does not now, much to the obvious chagrin of those who wanted Venus to be capable of sustaining life.

Many witnesses have said that they have seen discs taking up sea water or water from lakes. The inference is that the aliens are after deuterium (a heavy isotope of hydrogen) for their drive systems. However, as there are only about 160 parts per million of deuterium in the sea or any other natural water supply, the aliens would have to process vast quantities to get at the precious deuterium. It could be, of course, that they are after marine organisms or fish rather that isotopes of hydrogen – it could even be that they want to wash out the laboratories where they take apart cattle and other animals. Do not think that this is frivolous: in the absence of any proof, that possibility is as valid as any other.

The Earth teems with all sorts of animal and plant life. Could it be that the aliens are on a kind of protracted intergalactic research mission? Is Earth an oddity rather than the normal sort of planet you might come across in millenia of travels across the Universe? Is that why *they* seem to be staying so long – fifty years in modern times? What happened before that? The Earth's history is littered with legends and eyewitness accounts of strange beings from the skies. In some cases, these unearthly beings were even said to have assisted in human beings' development. Did we invent the wheel, or was it given to us? The electric generator? Telecommunications? The understanding of nuclear fission? The list is endless. Perhaps just key items were given to us and the inquisitive human mind did the rest – or perhaps we did it all ourselves, with no help from anyone else.

If you believe that the aliens have been observing the Earth for a very long time, ask yourself whether or not *you* would intervene to help in times of confusion and distress, especially if you *knew* the answer. How many times have you intervened to help your child when you have witnessed their frustration with a problem they cannot solve alone? Even more importantly, how many times have you protected your child from potential danger when you could see that they were unknowingly at risk? How many times have you corrected and steered your child on to the right path? Are we *someone's* children?

The picture becomes benign: the implicit love of one completely different race for another. Could *we* do that? Are *we* capable of such loving? We cannot even solve our own racial problems on Earth, let alone love an alien race from another star system! And why should *they* love us? What can *they* possibly get from such an insecure, violent, cruel and selfish creature as the human being? Or is the love that allegedly flows from *them* in an encounter a chimera, a construct brought about by their technology to assist *them* in their aims? Could it be that *they* only select and abduct caring, gentle folk? History tells us that that is not the case.

It has been noted that the greys in particular are clearly frightened of our physicality. Why hang around then? Even with all the technology at their disposal, our crude physical weapons are still lethal to *them*, and yet *they* persist, despite all the undoubted risks *they* take just in being on Earth. Why? Perhaps *they* know us better than we know ourselves. Are *they* interested more in people than in 'oddity' Earth? If *they* are, why do *they* still take cattle parts? Why haven't *they* openly made themselves visible to the nations of the world? Surely after all this time *they* would have contacted the UN and appeared on television across the globe? Why are governments still trying to cover up UFO events even at this late hour? Could it be that *they* are here to protect us against a natural catastrophe, like the great comet that is predicted to pass near (or hit) Earth around 2026AD? Could it be that *they* will evacuate the planet before disaster strikes in the form of the mythical 'Tyhon', before we are smashed and obliterated by 1,600km- (1,000 mile-) high waves as the comet's tail brushes by the Earth in a close encounter, or before we are crushed and frozen out by a nuclear-type winter if we take a direct hit? Maybe a direct hit will alter Earth's orbit, sending it into an eccentric elliptical path around the Sun, with the dire consequences for all life which that would entail.

Could it be that as part of their programme the aliens are breeding a new race from the remains of their own and from human stock, so that survival is assured after the unavoidable cometary disaster and destruction of humankind? Could it be that we are a zoo, owned by these alien creatures who visit and monitor us, performing operations on us as we do on our zoo animals? Maybe we are their experiment, planted on Earth countless millenia ago from another planet in some nameless star system *they* had discovered in their galactic travels? Or maybe we are a galactic tourist spot for the many alleged alien species visiting us?

THE ALIEN CAPABILITY

Let us compile a brief resumé of the overall picture. All the aliens visiting the Earth appear to possess technology that allows *them* to come and go as they please. *They* can neutralize or disrupt our electrical systems or installations at will. *They* can manipulate our time and our spatial reality. *They* can abduct us at will, sometimes taking a wife or husband from the same bed without waking the other partner, or a driver or a passenger from a car without disturbing the other occupants. *They* seem to have no respect for the physics which binds us to our reality, for *they* can reportedly materialize or de-materialize at will, travel through solid objects and sometimes take their victims with them, and they appear to have the ability to control our minds.

Some of *them* seem more interested in plant life than in humans. There is a presumed predilection towards cattle or animal experiments, where certain body parts are taken but the carcasses left behind. There is a disturbing trend towards human embryo experiments, with eggs and sperm being taken, and unexplained terminations of pregnancies without any miscarriage taking place. There are reported 'gifts' of alien technology in some kind of exchange with the US government. There are more reported sightings of strange aerial objects than ever before all over the world. There is an increasing acceptance by ordinary people of the existence of aliens and of humankind's helplessness in the face of *them*. Very slowly, an acceptance of the reality of an alien presence on Earth is growing. Whether this is an intergalactic or intra-dimensional reality, or whether *they* have always been here is immaterial – the fact is, *they* do seem to be here and in ever-increasing reported numbers. If *they* are intergalactic, is it possible that, relativistically speaking, they picked up our Hertzian chattering 'yesterday' and decided to come and take a look – because *they* thought that *they* were the only life in the Universe?

Our Universe is a very big place indeed, and even if our visitors have mastered interstellar travel *they* may not have thought to explore the distant fringes of our nebula where we live. Unless, of course, we alerted *them* to our presence. If our visitors are intra-dimensional, listening in to radio chit-chat does not really enter into the equation – they would already know that we were here. Could it be that we have received help from *them* throughout our evolutionary history? Are the early remains of ape man currently being found in the African alluviums the fossilized results of the aliens' genetic experiments to lift humankind out of the swamp? Sci-fi nonsense some would say – but is it? One of the main objections raised against proponents of the theory that interstellar visitors came to Earth in the distant past is the old chestnut that nothing has ever been found which proves that they have been here. Is it reasonable to assume that if anything was left on Earth it would be knowledge, not monuments? Knowledge lasts. Wait a minute! Perhaps *Homo sapiens* is the monument?

Perhaps the real truth is that *they* are both intergalactic *and* intra-dimensional. Perhaps the two definitions are not mutually exclusive. We know so little about our Universe and what we do know is limited by our insistence on trusting our methodologies of observation and the scientific paradigm of repeatable experimentation. It would be ironic if those splendid attributes cause us to be prisoners in our reality due to a lack of imagination. Perhaps one of the visitors' purposes is to make us aware of our potential by demonstrating their subtle technologies and what *they* can do. We have much to learn.

Think about how our societies function: how we get around, for instance. Think of the brutality of our machines, the violence that propels a train, a boat, a plane, and sits just beneath the surface waiting to ensnare and kill us when things go wrong. Our technologies are based on violent principles because we know that such physical systems work. They work wastefully and crudely, but they do work. While the technology works, we have no need to change things – that is, of course, until we have to.

Think about what the aliens have demonstrated to us with their technology. Aeroforms that defy all our laws of aerodynamics. Materialization and de-materialization, apparently at will – *they* are so obviously capable of disobeying the physical laws which limit us. *They* also seem to possess very advanced and powerful telepathic powers which render speech unnecessary. (Some observers have heard aliens communicating among themselves. *They* may, therefore, use telepathy as a means of communicating to us only because we may be incapable of comprehending their speech. It may also be a method of restricting information, i.e. *they* only tell us what *they* want us to know). *They* do not use oil or gas, rocket fuel or even electricity as we do, and yet *they* can do all these miraculous things. Is the heightened incidence of reported channelled information merely an innocent statistical by-product of more attention being focused on the phenomenon, thereby attracting more of the fringe and cult elements, or might it be a deliberate attempt by the visitors to reach as many people as possible by somehow beaming us information and, more particularly, the wisdom to use the information *they* give us? Free your imagination.

The aliens continually puzzle and tease us by mimicking our crude technologies, viz the many reported incidences of rocket-type exhaust flames and roaring sounds of propulsion. Occasionally physical traces are left behind which, when analysed, turn out to be conventional metallic substances commonly found on Earth – albeit sometimes in unusual combinations. Some are said to fly in pure magnesium discs, but the discs do not burn up through atmospheric friction caused by the tremendous speeds we witness on radar and with our eyes. Why not? Yet other examples are reported to be made from unusually light but indestructible metals which we cannot burn, cut or grind. Discs seem to weigh in at several tonnes, judging by the indentations seen in soil, yet they can make erratic manoeuvres which would destroy any of our aircraft, let alone our pilots.

The discs can be invisible to sight and (on occasion) radar, and they can be almost completely silent. They can hover motionlessly at treetop height in a roaring gale and then in a few seconds accelerate vertically to a become a pinprick or point of light in the sky. They have been known to leave 'angel hair', a peculiar gossamer-type substance which soon evaporates. Other trace deposits can be like sinter, soot, tar or bitumen, grey dust and myriad other substances. All of these are very much Earth-based products – why should they be?

It is difficult to see what purpose lies behind these relics of the aliens' presence unless it is either accidental, as a consequence or by-product of an abduction, or a method of psychological reinforcement – a trick used to increase the power of a suggestion. Such reinforcement could be used with confidence to ensure that the victim connects to the extreme strangeness of the encounter via the sight, smell or touch of familiar things that they know to exist, thereby anchoring the event to the victim's normal view of the world. Unless the system of reinforcement is present, it is possible that the victim could suffer a reality shift, resulting in lasting mental damage. The aliens may therefore be attempting to soften the strangeness of their presence and their approaches using this mechanism: firstly, to avoid blowing our minds, and secondly, perhaps, to preserve a serial victim's usefulness to their agenda.

The aliens' use of our crude technologies may be connected to a programme of familiarization, i.e. discovering what makes human society 'tick'. By mimicking our behaviour, *they* may be seeking finally to understand us, and perhaps in that process we may come to understand more about *them* – that is, as long as we know that this is their intention in such a mutually beneficial exchange. It has also to be said that, crude as it is, our technology works best in our reality. (Their technologies, therefore, logically work best in theirs.) Is this another reason for their mimicry?

Let's take a closer look at some of these rhetorical questions.

UNDER THE MICROSCOPE

I shall give you what no eye has seen and what no ear has heard and what no hand has touched and what has never occurred to the human mind.

Translated from *The Coptic Gospel of Thomas*
Nag Hammadi Library

AN ALIEN RESEARCH MISSION

The aliens reportedly visiting Earth seem to fall into two main groups. The first group, which probably make up the majority, behave in a tentative and sometimes quite fugitive manner, as if *they* do not want to be seen on their forays gathering plant specimens. This group seem to wear breathing equipment and are clearly operating in much the same way that human explorers might behave in an alien environment, that is to say with extreme caution. From their behaviour, it would appear that *they* are unaccustomed to our planet and its ecosystem. It is therefore quite likely that *they* are on some kind of scientific or exploratory mission, and although *they* are unlikely to require secret bases on Earth, they may keep returning until their mission is either fulfilled or amended in its scope or range. It is perhaps significant that the other main group of aliens do not appear to object to this arrangement, either for their own sake or possibly ours.

The second main group of aliens have (it seems) permanent bases on Earth. *They* appear to be dug in at several places around the world, such as Puerto Rico – possibly with the co-operation of host governments hoping to profit from the technology *they* bring. These aliens, the greys, seem to be operating as though *they* are here to stay. Their playful encounters with jet fighter planes could be more a demonstration of superiority than an act of friendly jousting. The greys also tend to be assertive, despite their reported 23kg (50lb) frame. *They* have other support individuals around *them* who are not of the same appearance but who take part in their activities and are usually seen by abductees. The greys seem to be more interested in our biology than our culture and are completely dismissive of our technology. It is therefore reasonable to consider that *they* are not tourists on an intergalactic sightseeing trip. While the greys could be on a scientific mission, *they* are more overt than the other group in their actions (could this be a result of their confidence being boosted by the ongoing co-operation of many of our world governments?).

Is Earth an oddity?

Yes, it most likely is, at least in the known Universe. It has only just become technically possible to see attendant planets around star systems, so it is difficult to be sure one way or the other. Nevertheless, a radioactive hot interior with so much water and a virtually 'plasticine' planet may be rarer than we imagine. Our Earth is clearly optimized for the kind of lifeforms we are familiar with. Or, to put it another way, the lifeforms are optimized for the planet. Some aliens (notably the greys) can breathe our atmosphere and are at home with our gravity, which should mean that *they* have a similar environment on their home planet or planetary base. It could also mean, of course, that *they* have adapted their physiology to our environment over time, but it is reasonable to hypothesize that *they* would not expend effort on physiological adaptation unless it were absolutely necessary. Physiological adaptation may take too long to accomplish and there may be risks for *them* – if *they* ever wanted to return to their home world.

An historical presence?

This is one of the strangest aspects of the subject, and one of the most inexplicable. Jacques Vallée may argue that the aliens have always been here. Fairies, spirits, the jinn, et al.... history is littered with legends and stories in all cultures. C. G. Jung postulated that the human psyche is imprinted with archetypes and these phenomena fall into such definitions. Erich von Daniken would say that various alien presences have guided our world and its peoples over millenia upon millenia, the inference being that we should not be where we are today without their guidance and inspiration – but to admit that would be to diminish the human spirit and intellect. If the aliens are intra-dimensional and can skip from their time to ours (via 'wormholes' or interstices?), it is possible that *they* have always been here. It is then necessary to explain why *they* are still carrying out what appear to be crude abductee experiments and why *they* reputedly continue to mutilate our cattle. Could it be necessary for their new quotas to learn how to do things? *They* may lack the ability to learn by example – which is, of course, in direct opposition to the human way of doing things. We set great store by teaching by example – this is the way we operate and in fact it is difficult to see how any other system of learning could work.

The way we run our societies, and indeed our world, is dependent on many factors, not least our lifespan. While the average Western male lifespan is increasing each year (with female lifespans generally marginally greater), it is still only around seventy years or so on average. Lifespans in the less developed world vary wildly, from under fifty years to exceptional longevity of over one hundred years recorded (but generally unsubstantiated). Better food, vitamins and health care all contribute to longer human lives. Imagine, therefore, how you would live your life and run your society if you knew you would live to be at least 400 years old and possibly older? That is the purported lifespan of a grey. Undoubtedly, one of the things you would immediately do would be to slow down or eliminate all the rushing around we do in order to pack as much as we can into our very short lives. It quickly becomes evident that a species which is so long-lived will have different priorities to a shorter-lived one. It also becomes apparent that a commensurately long-lived society would have much more continuity and therefore stability, but on the other hand it may risk atrophy and procrastination.

INTERVENTION

It has been asked why, if our visitors are so all-powerful, *they* don't intervene when disaster strikes us poor humans? The interesting thing here is that rather than interfere, *they* generally just observe; that is, unless *they* are threatened themselves. While there are reports of overt action being taken by discs, it is not a normal occurrence. True, jet fighter weapons-control panels have been neutralized before pilots can fire their guns and missiles, but passenger aircraft have not been downed by neutralized control systems even though passengers and crews have had front-seat views of an aerial encounter.

When the reactor at Chernobyl melted down, a disc was allegedly seen to be observing the proceedings. If there was a disc in attendance, the occupants must have known that something utterly disastrous was happening, but *they* did not do anything. The reader may be forgiven for asking themselves what *they* actually *could* do. This does not bode well for those who believe in the 'space brothers', who will come down from the skies to rescue the chosen ones at the appointed time before the Earth is destroyed.

When intervention does occur it is usually on a very personal level, rather than a grandiose gesture of omnipotence or demonstration of power. Individuals may be cured of cancers and some are given 'second sight'. The general long-term effects of contact are sometimes beneficial – although not always. The individual approach is interesting, inasmuch as it is the same technique that was used by some previous great world teachers and prophets. However, the aliens have nothing to prove. Their technology seems to allow *them* to do anything *they* wish, so why should *they* help just one insignificant example of the human race? The common man or woman as a singleton probably does not feature in their overall programme and one contact is generally all that it takes to do the trick. One answer seems be that it is pure serendipity on the part of the human being involved – right place, right time or, paradoxically, wrong place, wrong time.

There are a few people who are serial contactees, perhaps 'tagged' individuals. However, it is not clear that there is any defined programme or agenda at work in these cases, except that the aliens may find it more convenient to stick with one individual for some time if monitoring is the objective. In that event, 'tagging' may be suitable if long-term tracking is required. However, the aliens would have to be very careful indeed with their technology if *they* were not to be found out. Whitley Strieber sums it up nicely by stating that no one person is special and that we are all (that is, the whole human race) equal in importance. The conjecture that the aliens need repeat witnesses to publicize their presence is surely outmoded: *they* already have far too much publicity.

Whitley Strieber also maintains that we all receive the same 'jump' in awareness, rather like a huge rachet being turned up an additional, regular notch, thereby benefiting all the world at the same time. Unfortunately, some nations do not seem to be receiving the message of peace and wisdom, and everyone knows that ferocious conflicts are aflame all over the globe, with a corresponding terrible loss of life. It is more likely that the aliens' programme is one of familiarization and education rather than intervention in our affairs. Nevertheless, their real intention may be far from educational.

GREAT INVENTIONS

It is fun to speculate as to whether we have been helped along our evolutionary way by the visitors, from the claim of genetic intervention (Bob Lazar says he has seen documentation stating that human beings have been genetically altered some sixty-five times by alien intervention to get to our current stage), to the suspicions of many that some historical artefacts are rather more than they seem and that the great inventions or discoveries are not solely the work of humankind.

Nikola Tesla
It would be wrong, of course, to denigrate the human being's indomitable spirit and great achievements, sometimes through considerable effort and at great personal cost. Nevertheless, history does indicate that in some instances we did not do everything on our own and in fact received help from elsewhere.

Nikola Tesla always maintained that he spoke to aliens and that his great contribution of the AC electric motor was the result of a sudden visual inspiration. It was mainly as the result of this pronouncement that his scientific colleagues ostracized and eventually reviled him. In spite of this, Tesla was eventually credited with the discovery of the electromagnetic propagation effect. However, in contrast to Marconi, Tesla saw the propagation of electromagnetic energy as a way of distributing electrical energy rather than a means of global communication. While Marconi toiled with the problem of transmitting the human voice over distances using the scattering effect of electromagnetic radiation, Tesla grappled with his lifelong objective of transmitting free electrical energy to all by pumping vast electrical potentials into the ground in one location to have it extracted on the other side of the globe.

Tesla was a prolific inventor who was self-taught. That fact also did not go down very well with his contemporaries. He is supposed to have invented (or at least conceived of) potentially terrible weapons, as part of his somewhat modernistic view on preserving peace. He invented the useful 'Tesla coil', a circuitry which is designed to multiply its frequency to achieve very large potentials. He also invented a mechanical oscillator which he claimed could bring down tall structures, such as a skyscraper, with ease. He dabbled with so-called 'death rays' and an effect known as 'scalar vectoring', which could destroy structures at a great distance – in the early to mid-1920s such devices were topical subjects for discussion.

In 1924 it was claimed that one Harry Grindall Matthews, in London, had invented a 'diabolical ray'. He claimed that this ray could defeat a whole army at a distance and destroy aircraft or ships before they reached any coastline. It apparently had a range of about 6.4km (4 miles) and the maximum distance would be about 11.2–12.8km (7–8 miles). Tests had proven that the ray could stop the electrical systems of cars, and gunpowder had been ignited at a distance of about 11m (36ft). Other, rather more prosaic but nonetheless spectacular demonstrations showed that Matthews could electrocute mice, kill plants and light oil-lamp wicks, all by the beamed use of his ray.

Almost immediately after the news of the 'diabolical ray', a story emanated from Russia that an engineer called Grammachikoff had invented an electromagnetic ray which could destroy aeroplanes. As a result of a successful demonstration of the system at Podosinsky aerodrome close to Moscow, the authorities decided to go ahead with the construction of a number of stations around the capital and other sensitive areas. It was also decided to construct bigger and more powerful stations to disable large warships. The demonstration of the device was apparently so successful that the Commander of the Air Service decided that it would be prudent to cut back on the air fleet because they would not be required now that Russia had the 'diabolical ray'.

Tesla's 'death ray' experiments were well known to those who lived in Colorado Springs. Ground and air potentials were so fierce that sparks would shoot from anything that moved. When Tesla deliberately induced sparking, lightning flashes could be around 40m (130ft) long! He later bragged that his 'lightning transmitter' could destroy anything within a radius of some 320km (200 miles).

However, Tesla intimated that his system did not use any kind of ray but instead massive electrical potentials which would produce a similar effect to being struck by lightning.

Tesla continued to develop his idea of transmitting electrical energy worldwide using very low frequencies, while Marconi concentrated on the rather more useful yet less colourful idea of the radio. Tesla maintained that his transmitter would produce around one hundred million volts and up to a thousand amps – which equates to around a hundred billion watts! He considered that it would resonate at around 2MHz (the Earth's natural frequency is around 6MHz), producing an amount of energy equivalent to approximately ten megatons of TNT. He intended that all that energy could be focused on anything, with predictably devastating results.

Unfortunately – some might say fortunately – fate caught up with Tesla and he lost his backers and funding. He sold his patent rights for the AC motor and generators to George Westinghouse for what turned out to be a knock-down price. Tesla then faded into obscurity, destitute and with his achievements unrecognized.

However, there is a postscript. In the early 1900s Tesla was supposed to have concentrated on his plans for the transmission of energy using radio frequencies. It has been suggested that he made one last grand attempt to get his ideas recognized and to prove that they were workable. On 30 June 1908 a massive explosion rocked northern Siberia near Tunguska. It has been estimated that the damage caused by the explosion would be the equivalent of that inflicted by over thirty million tons of TNT – some explosion in 1908! Theories abound – from meteors or bolides (unlikely, as no definitive remnants or relics have been found), through interplanetary starships blowing up and the official version that a massive fragment of Encke's comet blew up high in the Earth's atmosphere (therefore no crater), to the substantial effects of Tesla's beamed weapon using scalar vectoring techniques.

There are some students of the subject who believe that Tesla may have conducted such an experiment in one last defiant act, in desperation for the recognition he deserved. Material is still available for those interested in the subject via the various Tesla societies and research organizations, who continue to experiment using his original patent drawings and blueprints (details are given in *Nikola Tesla's Long Range Weapons* by Oliver Nichelson and *Energy Unlimited* by Dale Pond – see Bibliography).

Whether or not Nikola Tesla somehow caused the explosion is open to conjecture, but the events at Tunguska are worth re-telling, as they remain of great interest. It was said that a fireball was initially sighted at dawn by caravans negotiating the Gobi desert. People in southern Russia claimed to have identified it as a tube-shaped object glowing with a blue/white light and trailing a multi-coloured scar across the sky. At around 07h17 the object exploded. A farmer sitting outside his house some 64km (40 miles) from the epicentre of the blast was burned by the heat of the flash and people were lifted off their feet by the force of the explosion. Houses were cracked and damaged. Near the epicentre, everything was devastated. Reindeer herds were obliterated, leaving only a charred and burned mess. The fire of the explosion could be seen at a distance of some 400km (250 miles) and thick black clouds rose 19km (12 miles) into the atmosphere. Shock waves from the blast travelled twice around the world and seismographs recorded tremors of earthquake proportions.

Professor Felix Zigel, an aerodynamics teacher at the Moscow Institute of Aviation, and geophysicist A. V. Zolotov maintained that the area of destruction was roughly triangular in shape, which is remarkable in itself. It seemed to the scientists that the explosive material may have somehow been caged before it detonated. There may also be some evidence to suggest that the object made a manoeuvre just before the detonation – perhaps it was a stricken spacecraft after all, and not a beamed Tesla weapon or a comet?

Compte de St Germain

The enigmatic Compte de St Germain who appeared in England in 1745 (and features in *We Are Not The First* By Andrew Tomas – see Bibliography) reputedly possessed the secret of eternal life and the legendary ability to manufacture gold from base metals – alchemy. (Alchemy also went by other names, one of which was a metaphysical invention and was supposedly the means of charting the progress of a soul from its beginning to the Godhead.) It was said that he could produce large flawless diamonds, which in time he would use to amuse King Louis XV at several lavish dinners, to the delight of his guests. Later, the king would have St Germain instruct him in the art of alchemy and place great assets at St Germain's disposal for that purpose. During the course of some of these experimental and instructional sessions it was reported that several new silk and wool dyes were invented by the king. The great philosopher Voltaire said that 'the Compte de St Germain was a man who never dies and knew everything', and Frederick the Great said that he was 'a man whom no one has been able to understand'. Even more remarkable were the stories about his longevity. A church register in Eckernforde, Germany, blandly states: 'Deceased on 27 February buried 2 March 1784 the so-called Compte de St Germain and Weldon. Further information unknown. Privately interred in this church.' Note that this register does not record the date of birth, nor any other identifying names. If the date of death is to be believed, according to other evidence it would have made St Germain 124 years old! Even more remarkable is corroborated evidence that St Germain was seen periodically across Europe in the years following. Stephanie-Félicité, the Comtesse de Genlis (1746–1830), an educationlist and prolific author as well as a pensioner of Napoleon I, recorded in her memoirs that she met St Germain in Venice in 1821. It has been estimated that he would then have been around 161 years old!

The only manuscript known to have been written by St Germain, *La Très Sainte Trinosophie*, is preserved at the Bibliotheque de Troyes in France. In it he writes:

The velocity with which we sped through space can be compared with naught but itself. In an instant, I had lost sight of the plains below. The Earth seemed to me only a vague cloud. I had been lifted to a tremendous height. For quite a long time I rolled through space. I saw globes revolve around me and earths gravitate at my feet.

The last known statement St Germain made was his comment that he was much needed in Constantinople, and then in England, to prepare two inventions which we would have in the next century: trains and steamboats. This was no ordinary man – perhaps he still lives?

Albert Einstein
Albert Einstein was born just before noon on a bright and beautiful Friday 14 March 1879. His father was Hermann Einstein, who ran an electrical engineering business, his mother Pauline, née Koch. Nothing in Einstein's early years gave any clues as to the great man's future genius. In fact, he did not do particularly well at school, mainly due to his inability to cope with the grinding regimentation of rote learning. His life was not financially secure until much later, at least until he had become the world-famous progenitor of the General

Relativity and later Special Relativity theories. Einstein's genius does not seem to have come from his particular mathematical ability, which in itself was prodigious, but rather from his remarkable insight into the nature of the physical world around him.

Einstein relates how he came about the principle of relativity: he imagined a man running beside a light beam and considered what that superfast man might see! He seemed to possess a power of visualization that few others of his time could imitate. Paradoxically, much of his theoretical work had to wait for the necessary complicated mathematical tools so that his theories could be proven, at least in mathematical terms. Even today, much of his work still remains to be proven experimentally.

From the seventeenth century right through to the explosion of the first atomic bomb, there was a creative surge of development and interest in scientific subjects that, arguably, cannot be equalled even today. Now, great scientific breakthroughs are only achieved by those with access to huge research budgets and the facilities to carry out the work. Unfortunately, egos and the lack of interdisciplinary exchange often get in the way of progress.

Einstein was a rarity in science at a time in history which was ripe for new ideas on the reality of the world and what made it tick. Einstein's genius lay in the way in which he could visualize accurately the reality of matter. However, he was not always right and he never believed in a Universe without order. While he is remembered for his remarkable and incisive work on General Relativity and, later, Special Relativity, he is most widely known for his famous expression $E=Mc^2$ – which unfortunately, considering that Einstein remained a pacifist all his life, is connected to nuclear fission and the atomic bomb. He did other very important work in the realm of thermodynamics and provided the intellectual keys with which to unlock many problem doors. His last years were occupied with trying to unite Maxwell's electromagnetic theory, relativity and gravitation into one Grand Unified Field theory, which would explain the Universe. Unfortunately, he never achieved his goal.

There is no indication that Albert Einstein ever talked to aliens, as Nikola Tesla claimed for himself. It is as well for Einstein that he did not, for had he openly made a statement to that effect it would not have been long before the scientific community buried him professionally, just as they did Tesla. However, it is interesting to speculate as to whether perhaps the great man could have tapped into a vast hidden sea of consciousness that is denied to the majority? Could he really have had the ability to visualize the miraculous world of particle physics and the mind-bending macrocosm of the Universe without help from somewhere or something and, even more remarkably, without the sophisticated tools we have today? His insight and accuracy remains, even today, totally astounding.

If a sea of consciousness does exist (see Appendix, page 247), perhaps humankind's next evolutionary burst will be linked to a new renaissance of our philosophical and spiritual ideologies in order to control safely the incredibly powerful science currently at our disposal. Our evolution seems to be linked more to our mental acumen and the way we use the fruits of our technologies than to the physical models of Charles Darwin and Alfred Russell Wallace. Unlike Darwin, Wallace believed in a system where spirit caused changes to occur in the world of gross matter. He also speculated that the Earth's evolution may be directed by outside intelligences which could interact with human beings. His rather heretical ideas – when compared to the rigidity of 'natural selection' – may explain his historical omission from his rightful co-credit with Darwin for the Theory of Evolution.

It is unfortunate that science seems to spawn the worst endeavours of humankind. There appears to be an inevitability about it. It is entirely possible that Einstein may have wrestled with his conscience in his later years, in the knowledge that his remarkable insight into the physical world and its mysteries may accidentally have led to innumerable deaths when his contributions to science were inevitably misused by those who followed him and the A-bombs were dropped on Japan in World War II. Even the rational mind of Einstein was apparently touched by the almost spiritual way in which he unravelled the mysteries of science when he wrote: 'The most beautiful and profound emotion we can experience is the sensation of the mystical. It is the sower of all true science.'

The transistor

Legends abound concerning interference by alien intelligence in human technical exploits. One of the strangest comparatively modern enigmas concerns the legend of the 'surface barrier transistor' (SBT). According to Preston B. Nichols and Peter Moon, authors of *Montauk Revisited – Adventures in Synchronicity* (see Bibliography), alien technology is traditionally filtered through the US Navy Research

Laboratory (remember that Bob Lazar was employed by the US Navy – see page 78) to the general scientific mainstream. The authors contend that alien help and instruction was given to Earth scientists in the development of the SBT as part of a contract with Harry Truman's government in exchange for the administration's undertaking that no more A-bombs would be made, tested or developed after the Japanese bombings. (Harry Truman just will not go away! Remember the now discredited MJ12 papers?)

The group of aliens assisting in that process were known as the 'K Group'. The story goes that the US went along with the exchange until the SBT was delivered and then reverted to A-bomb manufacture, to the extreme chagrin of the K Group. Fortunately for us, the K aliens were not hostile (although extremely naive) and left the scene (or the planet) in disgust. Quite why the aliens should assist in the development of one particular type of transistor is not clear (see Appendix, page 247). Maybe *they* needed it for their own purposes? However, *they* appear to be able to advance their technologies far faster than we can and it may well be that wires, solid state devices and other electronics recognizable to us are obsolete in their technology. In the light of Bob Lazar's revelations, it appears that this assumption may be correct. The interference idea is a strange one inasmuch as it seems that the aliens are changing our future by dabbling in our present. If our visitors are intra-dimensional or exist in *our* future, they are currently contravening virtually all the accepted physical laws that we hold dear – unless, of course, they know something about our physical world and its ultimate future that we do not.

DEMOGRAPHIC DESIGN OR JUST SERENDIPITY?

Their natures fiery are, and from above, and from gross bodies freed, divinely move.

Translation by Vergil of Henry Cornelius Agrippa's
Three Books of Occult Philosophy, 1531

VISIT TO EARTH

Approaching the solar system from deep space for the first time, you would be dazzled by the brilliance and ferocity of the nuclear furnace of the Sun. You would then notice the emptiness of space and tiny Mercury, until you saw the perfect brilliance of Venus shining with a pure whiteness to hurt your eyes. Looking out further, you would see another brilliant orb, as dazzling white as Venus but with a distinct bluish tinge – Earth.

Any visitors to our solar system would no doubt carry out a thorough survey of all the planets and planetary satellites. Scanning techniques could be used to detect and diagnose all radiated emissions. In that way, it would be possible to detect intelligent (and therefore, by definition, dangerous) lifeforms from a distance without alerting those lifeforms to your presence by sending probes to the surface. Having carried out your survey, you would be delighted to find an apparently highly formed, intelligent lifeform on the third planet from the star which, although potentially hostile to you and your environs, was to all intents and purposes a prisoner on its home planet. Given that there is not much hope of the lifeform becoming more civilized and less aggressive within the timeframe allotted to your expedition, you would make it a priority to ensure that the lifeform stayed planet-locked for as long as possible.

However, your expedition is not in the sector in great numbers, so you cannot really do much but exercise extreme stealth and caution in your examination of this intelligent but still quite primitive species – primitive in inasmuch as it seems to operate on several disparate levels with the paranoia and insecurity forgotten by your own species countless millenia ago. You might use the opportunity to get some essential supplies that you cannot produce yourself using your on-board systems. Your scientists would want to research the lifeforms on this strange 'plasticine' planet to the full, especially as you know that the planet can support your own physiology with little or no adaptation. In fact, your technology seems to work more efficiently here than it does on your own home planet. This will give you an edge if you are threatened by the indigenous lifeforms. You soon learn that you can only be harmed in exceptional circumstances, and broadly speaking you and your technology are quite superior to anything on this beautiful but violent planet.

Your commanders note that the scientific survey cannot produce any more meaningful knowledge concerning the indigenous species (including the intelligent bipeds) or the morphology of the planet. While your mission programmes continue, you are instructed to make direct contact with selected Earth powers. You select your contacts carefully by using the universal language of mathematics and mind-control methods – you understand them well. You convince your overawed hosts that you mean them no harm. You demonstrate your technology and, while it is understood by their scientists, it cannot possibly be duplicated with their current understanding of physical laws or without access to your technology. You give them gifts which will cost you nothing but which amuse and please your hosts. Your technology cannot be used against you because it has been designed to be non-offensive, for this is one of the major rules of your society; there has been no need to be offensive to anyone or anything for countless millenia. Nevertheless, you know that you are taking a big risk by letting these undeveloped, aggressive bipeds possess your technology.

There is, however, a cost to the bipeds. In return for your gifts, you want permanent bases on their planet. It is awkward and costly in time for you to maintain a presence outside the planet, even more so since your scientists have completely mastered Earth's environment, in which you can now function almost as well as you would on your home world. They have even fine-tuned your physiology. So, you set up your bases. Soon you run out of room. You need to expand. You are far more confident now and your scientists begin a hybridization programme to increase your own numbers. You have found a world that you like and one that suits your purpose, for it teems with life – such a rare commodity in the Universe – a world at the azimuth of its evolution.

You have adapted well and the native bipeds' planetary societal structures are so underdeveloped that they really present no threat at all. In any event, your cerebral-cortex control technologies will take care of any hostility towards you. You have cleverly arranged it so that your contact is with the most developed nation (even though you do not understand the concept of nationhood – you have no equivalent on your home world) and the rest of the biped populations know little or nothing of your existence, your programmes or your strengths. In return, you have promised your hosts rewards in the form of some of your out-of-date technology, which they believe will increase their knowledge and stature in their world.

There are rules, of course, and very occasionally they are broken, usually by your hosts, through either ignorance or fear. There have been times when you have had to demonstrate your superiority and, unfortunately, some bipeds have died. Eventually, over a period of many Earth years, you come to exercise a degree of control which affects the policy decisions of your hosts. You quickly realize that you have more power than you ever imagined. You have consolidated your position and your programmes are safe from any interference. After all, you made a decision to take on the best on this strange plasticine world. In that way, inferiors would be of little or no threat to you.

Crazy scenario? Perhaps. While the first part of the above narrative is conjecture, the rest is based on what is legend concerning the greys. It has been suggested that the US was originally not alone in being selected by the greys. It may well have been that all the so-called 'superpowers' were invited to participate in a programme of mutual self-help (more likely self-interest). The stakes must have been very high because the belief is that the only superpower now left in the contract is the US. There are those who think the original exchange between the greys and the superpowers occurred back in the early to mid-1970s and this eventually grew into the alleged current uneasy agreement. Some kind of serious disagreement is said to have occurred in or around 1978, which resulted in the USSR being kicked out of the programme. The legend maintains that this episode interrupted proceedings for a few years but things are now back on track.

There is usually no mention by the sources who leak the information of the UK or France, or any other emerging nuclear power such as India or China, being part of the privileged group. While it is reasonable to assume that the first two countries may have been invited to join this exclusive club, it is probably unlikely that emerging nuclear nations would be of interest to either the greys or the main players in the planned programme of exchange, mainly because the timetable of events would make their inclusion unnecessary. Presumably the greys could, if *they* wished, take what *they* want without permission from anyone. However, *they* would not get the co-operation *they* need and *they* would not be given safe havens from which to operate. It is logical to conclude that if such an agreement actually exists, the greys would have preferred contact with only one superpower. The US would then disseminate information (or technology) as it saw fit – or otherwise.

The legend continues. This cosy arrangement between the US and the rest of the so-called First World could have lasted for some time, at least up until the fragmentation of the USSR, who by this time had already been ejected from the programme. The UK no doubt would have been content to remain a poor relation of the US, willing to use leading-edge technologies but largely unwilling actually to become a member of the club in case the cost was too high – the price the US imposed of blanket security at the very highest level on a strict need-to-know and compartmentalized basis. Unfortunately, security lapses do occur from time to time – there simply cannot be any subject which can be considered an *absolute* secret because intelligence intercourse must, by definition, take place.

Let us suspend the legend now. As British citizens know to their cost, the UK does not have a Freedom of Information Act or a written Constitution. However, the US has immense experience in the security game and can therefore legitimately cloak itself in the respectable, government-sponsored process of slow information release to citizens who know which documents they want sight of (and can afford the fee) and all the appearance is of democracy at work. However, without this in the UK the government is under no obligation to release any pertinent data. Rumours or security leaks on the subject of aliens or UFOs are usually officially ignored or explained away quickly, dealt with by the establishment press as irrelevant and in a totally anodyne way – the doors close tightly shut. Such items go unnoticed by the vast majority of people and are often dismissed by those readers who do notice as novelty reportage.

When the UK eventually gets its own FOIA, its citizens will find themselves in the same rat's maze as the Americans. Information which remains outside the Act because disclosure may compromise national security will still be out of the grasp of the public and of researchers who cannot afford to buy the data blind, not knowing in advance what they are buying.

It may be significant that the US leads the world in the development of certain kinds of technology such as stealth systems. Note that the word 'weapon' is not used: stealth technology implies non-intervention or engagement with your adversary. The US has, of course, used that technology for offensive purposes – maybe the temptation was too great? As far as the USSR is concerned, historians of the future may record that it was unfortunate that the birth pangs of a new 'Soviet Union' came at such an inopportune time for them to benefit independently.

If the US is exchanging favours for alien technology, we should expect to see the emergence of new science, or old science rejuvenated. The rumoured mixed-power Stealth airplane, the almost noiseless, unmarked black helicopters, and the so-called 'anti-gravity room' at NASA could be manifestations of alien influence in American science. These technologies could, of course, be products of some very clever minds at the California Institute of Technology (CALTEC), NASA or other US organizations who have vast research budgets to spend on what are fringe or 'black' projects. What *is* certain is the enormous funding for projects which are not readily identifiable as discrete auditable programmes. It has been suggested that much of this funding is siphoned off from more visible government spending, even though the Star Wars programme has now officially been shelved.

It is interesting to consider that radar/sonar stealth technology is currently applied to anything that moves and remarkable whispering helicopters are being developed, in a world which is now generally concerned with the reduction of lethal armaments. While whispering helicopters probably have a viable commercial value, there is little to commend air transport that has a low radar signature or a ship with the ability to be invisible to someone else's probing instruments, unless it is from a military standpoint. Even the justification for military-piloted stealth systems diminishes when they are compared to satellite systems that can resolve land- or sea-based objects to a few metres from 500km (300 miles) up on the edge of space.

Is it reasonable that such technologies would be developed unless for commercial gain? The many small bushfire conflicts around the globe surely do not warrant the vast financial investments for this science required from tax revenues, which should be logically reserved for more serious humanitarian programmes? That the US *does* possess alien technology and benefits by its use is rumoured by several researchers. Witnesses like Bob Lazar and the researcher Bob Oeschler (recently back out of retirement) have indicated some areas where alien science may have been applied.

THE HOLLOW MOON

In 1980, Don Wilson published his *Secrets of our Spaceship Moon* (see Bibliography). The book presents a challenging theory put forward by two Russian scientists, Mikhail Vasin and Alexander Shcherbakov, and published in a Soviet government publication in 1970, that our Moon is a hollow world. There have been comments that the book was intended by the scientists as a 'send-up' aimed at their colleagues but, as we shall see, there may be compelling reasons why that is not the case.

It is generally understood by our science that hollow worlds do not, and cannot, exist naturally. Ergo, the Moon was hollowed out artificially by someone or something for a specific purpose. When I was at school, my teachers taught that our Moon was either captured by Earth as it wandered aimlessly through space, or that it was formed out of the Earth when the Earth itself came into being. Unfortunately, neither hypothesis agrees with what is now known about our lunar companion.

The Moon's orbit around the Earth is very close to being circular. If the Moon had been captured, it would be virtually impossible to establish such an orbit, which would instead be elliptical. The Moon's geology is quite different to that of Earth and appears to be composed largely of rare elements which are in relatively short supply on Earth such as titanium, zirconium, yttrium and beryllium. The strange thing is that the Moon gives the appearance of being an 'inside out' world: that is, you would expect to find the material that is on the outside of the Moon, on the inside! No amount of meteorite bombardment would cause such a rich mineral harvest.

NASA set up seismic experiments on the Moon and then crashed unwanted lunar-orbiting vehicles on to the surface, as part of the Apollo programme. The experiments showed that on impact the Moon actually rang like a bell for several minutes, with shock waves performing several circuits of the lunar surface. Another strange anomaly concerns the lunar craters. The lunar surface is riddled and pockmarked with thousands of these; however, it has been calculated that the depths of the craters should be at least four or five times their diameters – which they are not. Many craters are in fact *convex*, not concave as would be expected.

There is also the question of the famous lunar bulge. One would expect the Earth's irresistible gravitational pull to distort the side of the non-spinning Moon that perpetually faces us out of circular. However, the expected bulge faces away from Earth! Very strange.

Innumerable other anomalies are now known about our satellite, but most of the data is still held by NASA and has not been released to date. The continuing speculation with regard to UFOs being seen by all Apollo Mission astronauts and while engaged in activities on the lunar surface, may lend some circumstantial weight to the 'hollow Moon' theory. For centuries humankind has looked at the Moon and seen strange lights, clouds and movements on or above the surface. It is interesting to note that life on Earth might not be as it is today were it not for regular lunar tidal forces of just the right strength to cause the wave motion and internal tectonic stresses that keep the Earth geologically alive. There seems little reason for the Earth to have such a uniquely large satellite unless it has a specific purpose – perhaps to cause and perpetuate life on our world?

Is it therefore too much of a leap of faith to imagine that some civilization in the long-forgotten past somehow steered the Moon into a stable orbit around the Earth in order to control our planet's evolution? Are the coincidences of an almost circular orbit, just the right mass to control Earth tides, just the right distance from Earth to blot out the Sun almost perfectly when a solar eclipse occurs and so on, too much to consider as mere coincidences rather than design? Maybe some of our visitors do not have to come to Earth from any great distance: perhaps *they* have vast underground bases on the

Moon. Perhaps *they* have always lived on the Moon and just do their 'shopping' on Earth? Far-fetched? Science fiction? Even worse, just garbage? With the technology available to *them* it would not be impossible to create a living environment *anywhere* that would provide their minimum requirements.

The 'hollow Moon' theory remains a theory until someone proves otherwise or NASA decides to release all their information (which seems unlikely). That anomalies exist is true. The rest is either tantalizingly circumstantial or pure conjecture, but it is exciting to speculate that our supposedly sterile Moon may in fact be harbouring life beneath its scarred and pockmarked face.

THE GREY NATION

Even if the greys had no flying discs, no demonstrable technology and just appeared to their witnesses out of thin air, *they* would retain an attribute which denotes intelligence – organization. Some people say that *they* have a kind of 'hive' mentality: that is, if there are a number of *them* in one particular area, *they* all seem to react simultaneously to an unspoken command much in the same way, perhaps, as in our understanding a shoal of fish or a flock of birds may react to a stimulus. We do not possess this ability and it may be useful to explore briefly why that may be the case.

It is commonly thought that humankind's history is one of a predator. Predators need their own specific areas in which to hunt and are therefore territorial by nature in order to avoid unnecessary conflict and the risk of injury or even death at the hands of their competitors. It was natural that the territorially dominant primitive male would allow only his immediate family members to share his domain; all others would be attacked and ejected, unless there was a challenge by another to become pack leader.

This is thought to represent our history – in evolutionary terms, our recent history. However, if we had evolved differently – that is, *not* as solitary male individuals with a territory to protect – we might have adopted a more collective approach to our societies. The female of the species has always been protected, revered and even worshipped because of the power wielded by one particular dynasty over another through fecundity.

The greys may have already evolved out of these impediments or, possibly, never even suffered them in the first place. Our perception of *them* functioning as hive entities may be a misunderstanding of their social structure, i.e. their hierarchy and their ability to communicate telepathically. Another, less transparent reason for their overall behaviour in our presence could be a result of the fact that *they* may be totally genderless. Perhaps this is why *they* appear to be preoccupied with our own planet-wide evolutionary system through the general (but not universal) interaction of two sexes, whether in the animal or plant kingdoms. This aspect may be missing or lost from their own societal structures on their home world. Our system seems to provide far greater diversification of species and types within species in both the animal and plant kingdoms. In this regard, we are probably part of, in evolutionary terms, a very young eco-system.

The attraction of a living system of diversification like ours to such a solo system would be compelling indeed. It could provide the solo system with a way out of the evolutionary dead end in which it may now find itself. The speculation that the greys (assuming they had the choice) have traded their biological development for the technology *they* now possess may not be so far from the truth. That their technological achievements would be in great demand by our world at large, and the US in particular, would be credible. Moreover, the trade itself, based on biological exchange for advanced technology, remains credible even if it is not directly linked to the biology of the greys.

If the above scenario is anywhere near correct, it is clear that there would be limits to the scope of such exchanges, particularly when the greys (who apparently have none of our social awareness or constraints) begin to strain the contract by operating as free agents in the programme, totally unfettered by any of our conventions, of which *they* would have no conception or appreciation. Such an abundance of biological material on Earth, coupled with an awareness that the other party could do little or nothing to limit or prevent the advance of their programme, would undoubtedly lead *them* to act in this way. This point may already have been reached or passed.

Giving the US the odd disc or two, which cannot be duplicated and have a somewhat limited operational life, in exchange for a licence to pillage biological material – animal, vegetable and, more importantly, human – is not a good trade, and is one which would be condemned immediately by the rest of the world *if they knew about it*. The really valuable grey technology will never be handed over: the ability to manipulate matter at will, to materialize or dematerialize like a sophisticated hologram, to float in mid-air – those secrets will never be ours.

The aliens' attempts at mimicking our technology probably serve two main purposes. First, by mimicking *they* demonstrate their power and superiority over us; second, there is a message of empathy: '*We* can do this too, look at us.' Whether this message is fully appreciated by those humans who witness such events is open to question. However, the first message usually sinks in.

That *they* are fallible is proven: the many crashes of their craft reported around the world support this. Their view of death may not be the same as ours. Their reputed 400 years-plus lifespan must change their perception of time and of the ageing process that touches all matter. The risk of annihilation or death through accident or disease (if disease has not already been conquered by *them*) may still be respected. However, the general human belief in an afterlife, brought about through fear of death, could be missing from their culture if *they* are as technically advanced as *they* seem.

In his early contact with the greys, Whitley Strieber talks about 'soul eaters'. Despite what religion teaches us, the word 'soul' is a very subjective term. 'Soul' to one person may mean their spiritual self, to another their essential essence or life force. To a doctor or neurologist it may mean the electrical activity of the brain. Another person may consider their soul a mirror image or shadow of their crude physical body, which is vitalized by *prana*, the living force. Yet another may consider the soul a myth crafted to cajole the living into better ways and to reduce human beings' fear of death. Most of these views consider that the soul is indestructible and immutable, to be rejuvenated, revitalized and given a new physical form at some time in the future. Those who subscribe to the concept of reincarnation actually put a figure on the time between each 'death' and the next incarnation – around 50 years.

From the point of view of the greys, the US must have been clear winners in the selection process. A vast country, with the highest consumption of energy in the world and therefore, by definition, the greatest energy producer; a country which has a high degree of technical ability, putting men on the Moon (now, strangely, disputed by detailed analysis of photographic evidence) and launching probes to the deepest part of our solar system and beyond; and a country whose scientists would be mentally prepared to receive denizens from other worlds – they had, after all, rehearsed contact in numerous

1. *Temple of Machu Picchu, Peru. This Andean site was only discovered in 1911 by the American explorer Hiram Bingham and is believed to be the only Inca city to have survived the ravages of the Spanish occupancy.*

2. *Inca stonework, Peru. Note the remarkable fit of huge blocks of stone weighing several tonnes put into place by workers using crude tools and methods of construction.*

3. *Teotihuacan sculpture, Mexico. Just one example of the fantastic stone art made by the Toltecs who were thought to have a connection with the old Mayan Empire.*

4. *Major Donald Keyhoe, a one-time director of NICAP (National Investigations Committee on Aerial Phenomena). Keyhoe was removed from office in 1969 by John Acuff who was linked to the CIA.*

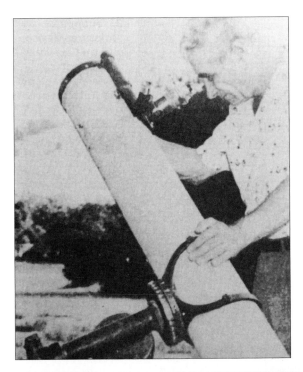

5. *The late George Adamski. Adamski was an early contactee who claimed he had first met aliens in the Californian desert in November 1952.*

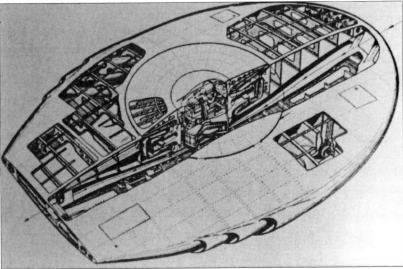

6. *The ill-fated AVRO Saucer Project; a US and Canadian government collaborative effort in the 1950s which did not mature.*

7. Author's impression of the Lazar 'Sports' model alien disk at S4 (Groom Lake, Nevada USA). Based on a design by John Andrews for the Testor Corp., USA.

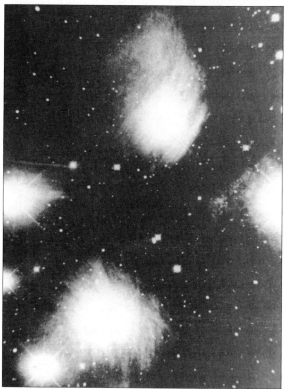

8. Nebulosity in the Pleiades. An astronomically young stellar cluster from which 'Billy' Meier claimed his beam ship occupants originated (see page 244).

9. *Still photo of an aerial object over Area 51 from an original video recording by Gary Schultz. Copyright Gary Schultz 1990.*

10. *The Andromeda nebula. This nebula is so distant that the light we see from it is around 750,000 years old. It would take around 50,000 years for light to travel from one side to the other.*

11. *Lockheed Martin M12 carrying D21-A, an uninhabited high-speed reconnaissance vehicle. This important project represented the thinking behind high-speed unmanned vehicles for intelligence gathering. (See page 247.) Copyright Lockheed Martin Aircraft Corporation.*

12. *Canadair CL227 'Sentinel'. Developed over twenty years ago for defensive maritime use, its strange shape might convince a distant observer that they were seeing a UFO. (See page 248.) Copyright Canadair.*

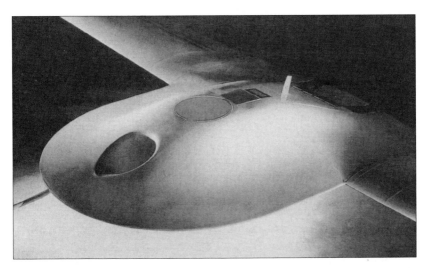

13. *Tier III Minus 'Darkstar'. This uninhabited vehicle represents state of the art reconnaissance technology. (See page 248.) Copyright Lockheed Martin/Boeing Aircraft Corporation.*

14. *The 'Predator'. This uninhabited vehicle is equipped with technical wizardry which gives it an optical range of some 2,100 kilometres (1,300 miles) at a cruising altitude of 7,600 metres (25,000 feet). This vehicle was successfully used in the Bosnian war. (See page 249.) Copyright General Atomics Aeronautical Systems Inc.*

15. *Lockheed Martin F22 'Raptor'. This aircraft represents the cutting edge of modern aviation technology and will be the replacement for the USAF's McDonnell Douglas F15 Eagle. (See page 249.) Copyright Lockheed Martin/Boeing Aircraft Corporation.*

16. *FOAS manned aircraft. An artist's impression of just one of the many conceptual proposals for a manned interdictor for service in the millennium and beyond. (See page 250.) Copyright MOD.*

Hollywood movies. This was also a country that had beamed out invitations from Earth for many years in the hope that contact could be established with other intelligent lifeforms, and one clearly equipped to take the lead in world affairs. Such a country would be the natural choice for contact, whether it is to be sustained or otherwise.

Continental Europe may have been rejected because of its lack of homogeneity and cohesion: myriad countries and cultures all operating at different speeds, no overall plan and far too many individual nations, some of which are barely civilized, almost continually at war and very unstable. The vast continent of Australia may also have been considered, but is too far away from anywhere and the Australians did not have the technology. Small island nations would have been considered and then rejected because of the lack of scope and support they could provide. No, the US was clearly the only runner in the race. They had the infrastructure and they were willing.

How would modern contact have been established in the first place? Conventional UFO history (if there is such a thing) would have us believe that contact was originally made by accident – *their* accident. If the Roswell incident actually took place and a grey was captured alive, it could follow that everything stemmed from the US government's contact with that one individual back in 1947. However, if there is any truth in the legends, affairs must have come a long way since the serious incident with the K Group to Bob Lazar's experience at S4 and the trades between the aliens and the US.

A superficial analysis of the human psyche might reveal that one of the major reactions in *Homo sapiens* facing contact with an advanced non-human species would be a willingness to generate a rash of altruism to other Earth nations and the urge to act as world peacemakers. This would demonstrate to our visitors that we are at least as civilized as *them*, if not quite so technologically advanced, and that we are really trying hard to become more civilized. It might be valid therefore to consider that the current altruistic preoccupation of the US with doing just that is not merely the usual cynical act of wooing the US voter.

NUTS AND BOLTS

It is just possible, although unlikely, that the greys have *always* been here on Earth. Perhaps *they* do not need flying discs, cubes, triangles or any other configuration in order to move around. However, their alleged ability to manipulate matter and ignore all the physical laws that bind us – by materializing and dematerializing, floating and passing through solid objects at will (sometimes taking their human subjects with *them*) – does not, rationally, indicate that *they* are some kind of transient phenomenon or that *they* operate in different dimensions or different resonances to our own. There could be at least three other reasons for that behaviour.

Firstly, it may be a predictable biological human reaction caused by their methods of interaction with us, i.e. the use of a mechanism to control our minds in order to achieve their objectives with the minimum fuss and, very importantly, with no physical risk to either party. Secondly, it could be just a manifestation of their technology, which we do not yet understand. Thirdly, it could be that *they* can operate in a sub-physical or metaphysical plane which is totally opaque to our senses or instruments.

However, the theory that the aliens can somehow manipulate gross matter could be viable in the light of current thinking at the leading edge of physics. There is some evidence to suggest that consciousness can influence what might be considered random events. In other words, the randomness of known random event generators – devices which may use unpredictable radioactive decay as a means of generating random events – can be influenced by the mind. A fascinating book which investigates this, and other, more profound thought experiments concerning the nature of matter, *The Holographic Universe* by Michael Talbot (see Bibliography), relates several experiments which tend to substantiate the idea that so-called randomness can be affected by our conscious minds. The jury remains firmly out.

For us, the argument of what reality is (or rather, what it is not), together with the bizarre notion that limitless numbers of independent realities can exist at the same time, tends to test our known, cautious physics to its limits. The dividing line between the physical laws according to which we believe we exist and philosophy is blurred when they meet and the argument becomes philosophical rather than scientific. (The 'Uncertainty Principle' perhaps gives us a clue to the real nature of matter: broadly, it states that there is a limit to the precision with which the position and momentum of a particle can be measured simultaneously). The act of observing a particle changes the nature of the particle. This is a little like checking your tyre pressures: purely by checking them, you are altering them. A few psi/bar extra to compensate for the release of the pressure pump gives no precise clue as to the *actual* pressure when you remove the pump. You are therefore unavoidably altering your tyre pressure merely by the act of checking it.

As everything is made from matter, it is reasonable to suspect that matter–matter interaction may be a normal event and non-interaction is not permitted, i.e. *that everything is permitted if it is not prohibited by nature*. Experiments in psychokinesis have proved that, under laboratory conditions, some people can successfully interact with matter to cause actual physical events. The manifestation of poltergeist happenings also tends to reinforce our suspicions that rather more is going on than we dare to admit.

If matter can react with matter implicitly and thereby cause physical changes, whether by the intervention of a living consciousness or, even more importantly, *without* that consciousness, this would forever change our view of life, our Universe and everything we have collectively struggled to learn over centuries of scientific evolution. Is it possible that a stone somehow *communicates* to a mountain that the plan is that the mountain must ultimately become a hill to fulfil its lifecycle according to the rules? Can our collective consciousness change world events? Can the power of prayer change society for better or worse? Are thoughts really *things* which can be tangible under certain circumstances? When we dream, are we constructing numbers of short- or even long-term realities? Is our Universe the product of a dream by an all-powerful creative master? Does all matter exist in a kind of universal patchwork – is it all connected somehow? Are we inevitably and inextricably linked to our sentient visitors – brothers to *them* by virtue of the fact that we share a consciousness, the act of being alive, even though we do not share their intellect nor *they* ours? It seems that consciousness, will, thought, call it what you like, *can* interact with gross matter and cause creative or destructive changes in the strange matrix we call space–time. Perhaps our visitors know the *real* nature of what we call 'reality' and have capitalized on their knowledge.

If a more conventional approach is required: if there were no structured craft and the illusion of every sighting, including radar signatures, were put on just for us as a piece of theatre, to delude us into thinking that *they* were as physical as us, it would no doubt still be necessary for some hardware to be involved to get our visitors into our dimension (and presumably keep *them* here for the required duration). No, the prospect of a purely dimensional spirit, a supernormal being, must be consigned to the reject bin. It is interesting to note that we only ever actually see our visitors at the time of our interaction with *them*, which implies that we are allowed to see *them* only with their permission. Therefore, *they* may just turn on their magic at the time of abduction. No one has ever seen *them* at play or while *they* are in normal mode, i.e. not kidnapping humans. No human has ever been treated as an equal to *them*. The kind of relationship *they* usually display is more formal and can be likened to a 'master and dog' relationship. It is interesting to note that the vast numbers of reported abductees are generally non-critical, ordinary folk who often have great difficulty in articulating and rationalizing their experience. Unfortunately, due to the large numbers of reported abductions it is difficult to see how any future reports can escape contamination.

Even disregarding their respective physical appearances, the 'Nordic' types of alien seem very different to the greys. The greys abduct and physically examine, mostly in the US, but the Nordics teach us spiritual ways and seem to have no interest in abducting anybody unless it is with the full agreement of their guest (not victim), in which case they take them for a quick spin around the solar system!

The descriptive name 'container' has been used by the greys to describe human beings. This somewhat dismissive terminology has never been explained (probably no one asked, or if they did they did not get an answer). To the more prosaic mind, the word 'container' may just indicate that we are made up of biological bits and pieces which seem to culminate somehow in a rather fragile living, thinking system. To the more spiritually minded, it could mean that we all have some kind of valuable life essence that can be reused over and over again to create new human beings, without the unnecessary personality getting in the way. However, for reasons of their own the greys do seem to value 'containers' and it would appear that *they* are genuinely fascinated by our existence. Perhaps sentient, carbon-based biological life is rarer than we think?

How MANY GREYS ARE ON EARTH?

The greys are the dominant type of visitor. Whether *they* answer to or co-operate with any of the other types of visitor such as the 'Nordics' or 'Giants' is unknown. However, their general approach is indicative of a group which is acting independently of any other alien types visiting Earth.

Someone once said that there are approximately four million aliens living on Earth at the present time – although quite how that census was carried out was not explained! As far as the greys are concerned, there could be any number spread around the world, most of whom are probably in North America (and the rest of the Western Hemisphere?) *They* all find it necessary to keep out of sight, which may be part of the deal with their hosts, and have underground or undersea installations. OK – so no one really knows because of the security clamp, but evidence from places like Puerto Rico, New Mexico, the European Atlantic seaboard and so on does indicate that there is considerable activity, some of which is blatantly overt on the part of our visitors.

According to legend, the other common type of alien – the so-called 'Nordic' – is not present in large numbers on Earth. *They* also appear to be of a more transient nature than the greys and flit in and out of our perception. However, *they* may be connected to the greys as *they* have purportedly been seen with *them*.

How long are they going to be here?

Most religions have timetables and prophecies of future events in order to consolidate the faith of followers. Unfortunately, prophecies do not always happen as forecast – as John the Baptist discovered several times. It is also confusing to the laity when it is discovered that the prophecies are couched in such ambiguous terms as to make them almost useless unless viewed in the most general terms and, it has to be said, usually for purposes other than the prophecy itself: the famous Quatrains of Nostradamus are a good example. The matter is further complicated by the fact that quite different calendars are used, depending on which group is doing the prophesying.

Despite all that, there have been, and are, many religious groups in the world who believe that the Julian date of 2000 will have some major significance in world affairs. Having said that, there were religious groups who thought that the last millenium would have turned out rather more eschatological than the event actually was. Prophecies vary from visions of apocalyptic destruction (note that some abductees have already been shown such visions), to a new dawn in Earth history where peace will last for thousands of years. There is little doubt that the greys might take full advantage of any fundamentalism to suit their mission objectives once *they* became tuned in to our human 'condition'. It is extremely likely that the year 2000 will come and go with the same rash of local world conflicts we have seen since the end of World War II. As we have not yet reached the point where the actual objectives of the greys are suggested or revealed, we must consider that the greys will continue to stay with us for the foreseeable future unless we kick *them* off the planet, which is highly improbable.

COULD WE CO-EXIST?

Probably not. Faced with the riches *they* possess in terms of technological, cultural, spiritual (?) or any other aspect of alien development, we should be at a distinct disadvantage. While *they* can breathe our air and cope with our gravity, there is some evidence to suggest that *they* cannot, at least at the present time, cope with our food. It has been suggested that this is due to their biology – an atrophied digestive system and different metabolism (if these fantastic reports have any grain of truth).

Some witnesses, like Debbie Tomey/Kathie Davis (see page 38), maintain ongoing contact from the age of six throughout adult life – Debbie produced nine hybrid babies and also suffered abductions of her own children (the full story is told in *Intruders – The Incredible Visitations at Copley Woods* by Budd Hopkins – see Bibliography). It is logical that if nine or more hybrid babies can be produced from one woman with no apparent physical diminution or effect, it would not take too long in a programme of mass hybridization to produce enough 'new' people to make a difference in our world – if the hybrids stay on Earth.

The question of coexistence would then take on a new meaning. We would no longer need to live with any *obviously* alien life: we would be coexisting with hybridized life, with the dubious benefit that the hybridized human would *look* almost exactly like us. The greys would, of course, still be in evidence. However, if it is assumed that the ascendency of the hybrids would be linked to the grey nation, who would no doubt be calling the shots, it becomes difficult to see what position of influence standard human beings would hold or purpose they could fulfil, unless it were to serve the hybrids, who in turn serve their masters.

Over a period of time, possibly a very long time, Earth nations would tend to diminish, even if the hybridization programme had other, less oblique objectives. A one-nation Earth might have obvious advantages for the greys, but for us there would be the disadvantage of a new level of mediocrity and stagnation as a distinct species. The greys might also be capitalizing on the splendid ability of human beings to accept and adapt to different races in their midst.

IMBALANCE

If the legend is true, the current state of affairs – with the US taking the lead in alien liaison – will eventually cause a terrible imbalance in world systems. We just cannot have one half of the world rushing into the technological future, with the other half peaking in development at subsistence farming levels and with no hope of benefiting from the new technology. There is no evidence to suggest that the grey nation, or any other alien group, would intervene *en masse* in world affairs if nuclear weapons were used against the wealthy nations by the poorer Third World countries as a means of economic leverage or revenge. That scenario is not improbable if one considers the potential prize of world domination that any politically energetic, ambitious group could envisage as a result of providing the Third World with the means to throw off its shackles. According to the legend, there seems to be no evidence to suggest that the greys have any insightful plan or goodwill towards the underdeveloped world.

As far as the developed world is concerned, there have been reports of 'fly-bys' and some speculation that electronic interference with control computers for nuclear missiles in their underground silos has occurred in the past. The interference could have been limited to a testing programme by our visitors to see how far *they* could go without actually firing the missiles or damaging the equipment, or it could have been a rather capricious and somewhat silly attempt at demonstrating their technical superiority over us – do we need any *more* evidence? This is an extremely dangerous game to play – unless it is a practical demonstration to prove that *even our most powerful weapons cannot touch them*. If that is the message, it tends to indicate that *they* are not at all concerned whether or not *we destroy ourselves*.

One is forced to the conclusion that 'space brothers' are not at work here. If the visitors buzz nuclear missile silos, provide specific advanced technology to the US, carry on kidnapping us without permission and continue to act as though we are collective spectators in an event in which we should be players, one cannot help but get the feeling that all is not well. That is not the kind of behaviour you expect from friends who care about you.

THE COVER-UP

Many people have speculated on the reason for the various governmental cover-ups around the world. The usual reasons given are that if the truth were known societies would collapse, governments would fall because of their embarrassment at being impotent in the face of the alien presence, religions would collapse and it would be Armageddon time for all.

Let's examine these claims. Embarrassment at being impotent in the face of the aliens is a situation governments have 'enjoyed' for a considerable time – so nothing new here. Societies and nations would not collapse, because societies do not *know* they are being infiltrated, therefore there could be no panic. By the time panic does occur, it will be too late to do anything about it anyway. The only religions that would collapse would be the ones with no flexibility or theology to support the idea that God, after all, made everything. All other religions would embrace a new lifeform as being part of the Grand Design: the Roman Catholic church is already on record as stating just that.

The common belief that we would face some kind of Armageddon is a result of the misconception that a sudden announcement would be made simultaneously to the populations of the world in a kind of *War of the Worlds* scenario (see page 181). It is very unlikely to happen in that way. You cannot conquer billions of people – people who would be more than willing to unleash aggression on invaders – by force, without causing a great deal of destruction all round, so this probably does not enter into the aliens' programme. There may also be considerable risk to the invaders, even though *they* possess superior technology. Nuclear fall-out would spoil the environment for *them*, thereby rendering the whole exercise futile. We should not, therefore, expect sudden announcements.

That some strategic advantage may be given to Earth nations who are blessed with inside information is in no doubt. That such inside information would make that Earth nation the most powerful is also in no doubt. However, the scale of imbalance caused by this would be extremely dangerous and must be addressed if disaster is not to ensue. The *type* of technical advantage should also be examined. If there is a choice of advantage, it should reflect a non-aggressive system, i.e. one that cannot be used for war. That may be extremely difficult, as no choice may be readily available and virtually any technology can be adapted for aggressive use.

It is perhaps timely to mention that currently the most overt use of supposed alien science allegedly involves secret use of their vehicles and possibly stripped-out power units in conventional jet aircraft for a mixed propulsion system... or so the legend says. If such an aircraft exists, it would need to be *extremely* reliable. The use of this particular technology tends to substantiate the contention regarding a non-aggressive alien artefact: gravity wave amplifiers do not, in themselves, pose any serious overt threat to world security through aggression. However, a real threat would be posed to the economies of societies (as they stand) if the amplifiers were manufactured by just one Earth nation who possessed the secret of their construction. Fortunately, the visitors have been wise enough to give us limited fuel supplies for that technology – fuel which, according to Bob Lazar, cannot be synthesized on Earth. This is very fortunate indeed. However, if the discovery of fibre-optic cables at the Roswell crash site has any substance in it, the same cannot be claimed, for fibre optics has not only revolutionized communications, it has also greatly enhanced our ability to create more robust and reliable weapons systems.

Bob Lazar's contention that the US possesses the secret of a devastatingly powerful anti-matter bomb as a spin-off of the gravity wave engine may have substance, but the question it is then difficult to answer is: why on Earth would anyone want such a device – a device that could devastate half the world in one go? There is no benefit in possessing such a destructive weapon unless one morning the world collectively wakes up to madness and decides it is a good day for mass suicide! If the US *does* possess such a terrible secret, it must always remain a secret. The awesome responsibility of such knowledge poses a further insurmountable problem for those who would use a gravity wave engine for economic advantage.

The legend continues: the cover-up may, of course, be part of *the deal*. It may include a clause which states that in order for exchanges to be made between the hosts and the visitors, the hosts must first demonstrate their ability to ensure that a suitably stable and peaceful world environment exists. The attitude and approach of the US over the past few years has tended to be less hawkish and more conciliatory towards its adversaries, as well as being very receptive to peace brokerage, and this tends to lend weight to the theory of conditional co-operation. That the US is very concerned to maintain a tight grip on its knowledge base is presumed correct. That the US would not want to share the alien secrets (or cannot – which amounts to the same thing) is also presumed correct.

If the contention that the deal struck between the greys and the US is exclusive is correct, the exclusivity will probably be a codicil to a far wider agreement and a prime condition laid down by the greys before any co-operation could take place. That the US/grey relationship has evolved considerably since first contact was made, even after a couple of false starts, may be correct. That few other governments are let into the innermost secrets may be correct. That even these few governments are not kept up to date on current developments may be correct. That the rulers and technocrats of the US – as sole representatives of *Homo sapiens* at large and with all the awesome responsibility that that entails – are the main contacts for the greys in this world may therefore also be frighteningly correct.

The cover-up appears to protect both host and visitor. However, whether it benefits both host and visitor in equal proportions is open to doubt. Could it merely be in place to protect the sensibilities of people and nations who would be frightened or emotionally threatened by the thought of alien lifeforms visiting Earth, let alone staying here? While the deal would perform such a valuable and necessary function, it is obvious that the cover-up cannot last forever, because of the nature and general conduct of the visitors together with the increasing number of disclosures from 'deep throat' sources. All this indicates that *the cover-up may no longer be part of the deal*. Things may have changed – perhaps irrevocably.

COLLABORATION
AND
CONSPIRACY

God is in the lowest effects as well as in the highest causes.

William Blake

There is little evidence to suggest that either the UK, any other European country or any of the Commonwealth countries has deliberately collaborated with alien entities. However, that does not mean that collaboration has not occurred. What it does suggest is that no one has been found out. If you believe that collaboration includes the poorly understood paranormal phenomenon of 'channelling', then according to some researchers the Nazis may have been guilty.

ANATOMY OF COLLABORATION

Collaboration means different things to different people. In the most commonly understood terms, it means that co-operating with a superior adversary is a far better way of staying healthy than opposing them. In other words, collaboration may be the only method available for survival. In return, there may even be the occasional reward from a grateful master. Collaboration, then, is generally considered to be a totally pragmatic approach to a situation which is somewhat less than ideal for the subordinate. The master always has the upper hand.

In order to keep the collaborative regime intact, certain requirements must be met. To protect secrets, less involved individuals (who have no real knowledge of the scope and quality of the collaboration) may need to be threatened, cajoled or rewarded, as necessary, on a regular basis so that the balance of master and subordinate is maintained. Occasionally, greater sanctions may have to be applied and an example made of an individual, and to achieve that the necessary infrastructure would need to be in place. The well-tried 'need to know' system would be in evidence, with individuals limited in knowledge to their own compartmentalized systems. No individual should know more than any other individual if the structure is to survive; some senior subordinate would know the 'big picture' otherwise there could be no cohesion or synergy in the partnership. Finally, it is worth noting that collaboration almost always works best when there is a continual presence of fear and threat.

Actual policy, administration and direction is not the meat of the collaborator: it is the responsibility of the superior partner to hand down such matters and to maintain control. If control is left to chance or unattended for even a short time, the subordinate may see an opportunity to gain an advantage, if he is bold enough. The collaborator does, of course, stand squarely between the master who dominates him and his non-collaborating peers, who may simultaneously despise and respect him, as might a caged wolf who both fears and hates his keeper at one and the same time.

Collaboration breeds conspiracy and such examples are legion in the ordinary affairs of human beings, usually in the name of national interest. Play and counter-play form the subject of most spy novels, but for our purposes we should try to focus on the alleged conspiracies forced on the US by their continued alleged collaboration with the greys. Of course, as with most other countries, there is probably a great deal of conspiratorial effort made in the legitimate defence business of the US. The web of intrigue, disinformation, false leads, denials and lies is woven into an incredibly tangled fabric in which the real truths are hidden. In every effective lie there is a small element of truth.

MORE DISINFORMATION

A t this point, let us briefly review the overall UFO phenomenon once again before continuing. UFOs have been seen for centuries, at least for as long as human history, if we are to believe ancient chroniclers. The modern wave of sightings started in 1947 courtesy of Kenneth Arnold, when he saw several flying wing-type objects skimming over the Cascade Mountains in the US. Since then, sightings all over the world have stepped up in intensity. Stories of crashed discs, and recovery of them and their crews, are legion. Tales of abductions, crop circles, cattle mutilations and genetic experiments are widespread. Ask yourself why a government would choose to use such outlandish and unbelievable events to cover-up the development of new aerospace products? Even the most secret of secrets surely does not warrant the use of such an exotic and obviously weird excuse, especially when an increasing number of witnesses are coming forward who maintain that they have seen what they believe to be alien technology at government installations.

It is reasonable to assume that some disinformation is spread in order to protect the status quo. Whether that disinformation is used to protect alien or conventional technology is immaterial, although it is reasonable to consider that absolute emphasis would be placed on secrecy concerning alien science. If you could keep the lid on this for long enough, it might even become integrated into your own programmes in time. All you do is let your science catch up with the new science – easy. Unless, of course, someone spills the beans.

Misinformation programmes may be endemic in both governments and the commercial world, in order to protect legitimately very sensitive or valuable information. What better way to spread misinformation and rumour than by using the various UFO clubs and enthusiast groups, consisting as they do of a usually willing audience who desperately *want* to believe? You might start by mixing an innocuous but tempting morsel of fact with a lot of fiction. These would be facts which are not in themselves injurious to your overall project and which, on their own, mean very little to anyone, for instance: 'It has been shown that discs emit microwave radiation' (possible) or 'The discs use anti-gravity to fly' (probably false). There is a risk that some believers will become addicted to your misinformation. Your target's belief system is augmented by the stories you spread and the data becomes a cocktail of truths, half-truths and lies, which eventually cannot be separated out by those who are fed this intoxicating diet.

Unfortunately, the road is littered with the casualties of such subterfuge. In his book *Revelations – Alien Contact and Human Deception* (see Bibliography), Jacques Vallée relates the affair of Paul Bennewitz, a respected physicist. Vallée maintains that Bennewitz was influenced by a woman who maintained she was an abductee. During several unreliable hypnotic regression sessions she had stated that she had been taken to an underground base and seen pieces of human flesh floating in vats of some kind of fluid. She was apparently also told that US scientists and greys were working together under New Mexican soil.

Some time later, Bennewitz allegedly rigged up electronic surveillance equipment to see if his witness was affected by outside influences. The detection of strange electronic signals spurred him on to discover more 'evidence' in the New Mexico area, which he apparently took to be where UFOs had crash landed. He was visited by many researchers over a period of time – the list includes John Lear, Jim McCampbell, William Moore and Linda Howe, among others – and confided to them that he had seen UFOs swooping down from the sky to abduct motorists. Bennewitz gradually became more and more obsessed with the subject of UFOs.

Jacques Vallée met with William Moore at a MUFON (Mutual UFO Network Inc) conference in July 1989 and Moore admitted that Paul Bennewitz had been the subject of a deliberate disinformation campaign by the US government. The story goes that Bennewitz had accidentally stumbled into a secret USAF project which had nothing to do with UFOs. He was apparently asked by security personnel to stop monitoring their tests. The harder security tried, the more effort Bennewitz put into discovering the secret, as by now he was convinced that it had something to do with UFOs. Moore said that the security personnel then decided that the only way to get Bennewitz to stop monitoring their project was to go along with his fantasy and convince him that aliens were everywhere. Helicopters on exercises then became marauding UFOs taking up people from the night. Bennewitz was thus discredited as a reliable source of information, thereby neatly relieving the USAF of any further inconvenience.

William Moore's breathtaking claim raises several very important issues. Firstly, why was such a high-handed and risky method used to discredit a tenacious and intelligent researcher? Secondly, why use a sledge hammer to crack a nut? Surely there were other methods available, which might even have included getting Bennewitz involved in the project, albeit in a minor way. Thirdly, the establishment never owns up to shady dealings and whistle blowers are either not taken seriously, discredited or, in the worst cases, never seen again. Fourthly, and most importantly, what are the ramifications for the US

Constitution? These questions tend to indicate that there is far more to this particular incident than meets the eye. Is it really credible that anyone could go public like Moore did and not expose themselves to all kinds of problems – none of which could be considered healthy options?

Jacques Vallée then tells us about astronomer and writer Morris K. Jessup, who apparently committed suicide by carbon monoxide poisoning on 20 April 1959. Jessup had written several books on the UFO phenomenon and reputedly was led to believe in government cover-ups, experiments in advanced physics and two races of ET, the LMs (friendly) and the SMs (not friendly). He also corresponded with a somewhat mysterious individual named Carlos Allende. Jessup had studied and lectured in astronomy at the University of Mexico. Upon becoming interested in the UFO phenomenon he wrote a book, *The Case for the UFO*. Apparently, a copy turned up completely unsolicited at the Office of Naval Research in Washington. The book was defaced by some writing in the margins, which indicated to the reader that the secrets of UFOs were solved once and for all. Jessup was invited to Washington to see this defaced copy of his book and was amazed at the apparent total familiarity of the writer with the subject. Jessup's interest was further fuelled by references to the alleged disappearance of the *USS Eldridge* (the so-called Philadelphia Experiment). Probably from the time that Jessup saw the writings in the margins of his book he was hooked. The abortive search to find the mysterious Carlos Allende drove Jessup to a state of obsessive desperation. Probably in a fit of the deepest despair and frustration, he had finally decided that there was only one way out of the tangled web that ensnared him.

Both these tragic cases probably describe authorized mis/disinformation at work and the real possibilities of an outcome where some individuals are driven to mental instability while others may not even fare that well. To recap: collaboration leads to conspiracy, conspiracy to disinformation and the spread of ridicule to discredit and destroy the credibility of researchers and prevent them from continuing their investigations. Presumably, if none of that works a convenient 'accident' is arranged.

In the past few years two new phenomena have hit the headlines – crop circles and cattle mutilations. While there appears to be no collective name for those who study mutilated animals, crop circle investigators are called, euphemistically, 'cerealogists', even though some circles appear in grass rather than cornfields. Misinformation processes may not be at work here, but there may be reason to suspect a programme of ridicule, especially with the admission of 'Doug and Dave's ' circular (hoax) exploits. This does not mean, of course, that there is not a real phenomenon to investigate, merely that the subject matter tends to be less serious or sensational than, say, a vat full of human body parts!

The cattle or other animal mutilations are another matter entirely and are taken very seriously by the owners of beasts that are killed and by those who investigate the killings. However, one cannot immediately see how these two subjects are acted upon either by using misinformation tactics or by singling out specific targets for ridicule and thereby discrediting the researchers. Theories as to how crop circles occur range from swirl vortices (a largely unknown meteorological event) to mischievous 'lights' seen working over(?) cornfields. Crop circle messages seem to be becoming more cryptic in content as we approach the millenium. Perhaps our visitors are not always culpable.

Let's consider why and what the US may be covering up. If we start backwards and believe everything that Bob Lazar says, it would not be difficult to comprehend the US government's reason for such a cover-up and their paranoia in its execution. It is, of course, also entirely reasonable to think that there is a more prosaic reason.

LEGITIMATE REASONS FOR COVER-UP

The much-vaunted Star Wars project initiated many radical ideas for the defence of the US. Stealth technology using radar-absorbent or non-reflective surfaces, together with low radar signature shapes and reduced-temperature exhaust emissions, was developed for aircraft and ships. Other (at least theoretical) projects included the investigation into and development of exotic high-energy particle beams that could blast an enemy's nuclear warheads in space before they re-entered the Earth's atmosphere and before they had a chance to range on their target. The US allegedly experimented quite successfully with programmes of physic remote viewing, telepathy and so on before embarking on the Star Wars initiative, so is it unrealistic to think that there was a great burgeoning of new ideas and theories to test around the time at which all this was going on – at least until the money ran out?

There is no reason to suspect that the stealth applications owe their existence to alien technology – radar-absorbent coatings and angular features are not typical of alien disc attributes. The reputed ability of the top secret Project Aurora to travel at over eight times the speed of sound at enormous heights on the edge of space also probably has nothing to do with alien technology, because it still requires airfoil lifting surfaces to operate. However, the rumour that there is a stealth bomber that uses a *mixed* power source, which enables it to hover as well as operate more conventionally, may indicate use of an element of technology which is not based on known mainstream science (more misinformation?).

That the US would want to keep their conventional science-based technology safe from prying eyes and cameras is entirely reasonable. Most nations with the wherewithal to produce such science would do the same, especially if the spin-off from the development of defence systems eventually opened the door to lucrative commercial markets which served to underwrite the prosperity of the nation for years to come. The introduction of radical new technologies for the military to be used later commercially is also reasonable: many of the technologies and lessons learned by NASA have other applications than that for which they were first intended.

The US intelligence community would obviously be aware of the suggestibility of the numerous UFO cult groups and clubs all over the country. If William Moore is to be believed, those groups have been (and probably still are) used to spread rumours, gossip and fantastic lies to mask the real truth of US progress in conventional science projects. However, consider for a moment: if the stories and rumours of UFOs and other otherworldly phenomena were disseminated by government agencies in order to discourage and confuse unwelcome investigators, the rumours would have to be increased not only in number but also in subject importance just to maintain the same level of security. In other words, the recipients of these rumours and stories would have become addicted and would therefore need greater and greater doses of disinformation to satisfy their hunger for sensationalism and excitement as they searched continually for the next exposé. Eventually the whole thing would get totally out of hand – and it quite often does.

That is one scenario. But remember: to work properly, disinformation must include an element of truth, but not so much as to let the whole cat out of the bag. You really have to make up your mind as to whether the chain of events described above is feasible, or whether more is at stake than the protection of the normal business of US national interest. An important feature which must not be overlooked is that high-profile official attempts to understand what was flying in US air space (Project Blue Book) were undertaken in the days and years following the 1947 watershed. Is it possible that disinformation, misinformation and rumour have now evolved into an unholy marriage of *legitimate* secret protection and *not so legitimate* protection of secrets that the world now has a right to know?

AN INTELLIGENCE CASUALTY

While there have been casualties among the UFO community, there have also been many casualties within the agencies on the opposite side of the fence. James V. Forrestal may have been one of the first high-profile people to suffer as a result of a conspiracy plot within the highest echelons of the Harry Truman administration.

James Forrestal had been the Secretary of State for the US Navy under Harry Truman's government. When the new unified command of the US armed services came into being in July 1947, he became Secretary of State for Defense. Coincidentally, July 1947 was the eventful month of the so-called Roswell incident; Forrestal must also have presided over other incidents, including the Mantell case (see page 32). In an expansive article by Dennis Stacey, editor of the *MUFON Journal US*, which was carried in *Flying Saucer Review* Vol 38, No 3, it is claimed that Forrestal was MJ3 on the infamous MJ12 Panel, which was allegedly commissioned by Harry Truman to look into the UFO problem and report back to him to ascertain what the phenomenon actually was and whether it was a threat to US security.

It was said that Forrestal was a very hardworking and conscientious Secretary of State who took his job extremely seriously. As Secretary of State for the US Navy, he had overseen the build-up of the most powerful navy in the world at that time. Navy ships ran on oil, and it is claimed that he was opposed to the creation of an independent Zionist state in the Middle East in case the US upset the oil-producing Arab nations. It is also claimed that he opposed the unification of command over the US armed services, mainly because it would dilute the independence of the US Navy. However, Truman elevated Forrestal to the new position of Secretary of State for Defense and it is said that he then softened his views on unification.

Allegedly, Forrestal's private life was also fraught with problems with his wife Josephine and his personal guilt at having left the Roman Catholic church. In-fighting and rivalries in the Truman cabinet took their toll as well. Apparently overlooked after years of loyal service, and worn out by the dramas in his personal life and within the government, Forrestal started to fall apart. He became unconcerned about his appearance and was often vague when asked to make decisions. Finally, Truman was forced to ask for Forrestal's official resignation and in March 1949 he left the government.

From then on things did not go well for Forrestal. In his paranoia, he imagined that his home was wired for sound, people were spying on him and somebody was out to get him. Eventually, he went to see an eminent psychiatrist, Dr William Menninger, who had been invited to Washington three months earlier to discuss combat fatigue, when Forrestal was interviewed to discuss the matter. Dennis Stacey emphasizes the fact that at that time the doctor found no trace of combat fatigue or any other mental or emotional abnormality in Forrestal. However, three months later the doctor revealed that Forrestal was now suffering from 'severe reactive depression' as a result of anxiety and paranoia. This condition was so severe that it was considered that Forrestal might commit suicide.

It was decided that hospitalization was the only answer and Forrestal was admitted to the Bethesda Naval Hospital in Maryland. Although sedated while in air transit, his paranoia flared and he convinced himself that his enemies were out to get him. On reaching Bethesda, he told his aides that he did not expect to leave the place alive. It is said that the White House eventually took charge of the treatment of Forrestal and the PR surrounding his internment.

Inexplicably, for a man with an illness that inevitably led to a suicidal disposition, he was moved to a VIP suite on the sixteenth floor of the hospital on the instructions of (quote) 'the people downtown'. Heavy screens were placed over the windows to prevent Forrestal from jumping. Interestingly, Forrestal admitted that while he thought that he was capable of killing himself using pills or a noose, he did not care for heights.

Apparently, one day Forrestal sent for Rear Admiral Souers to sweep his rooms for bugging devices, which Souers did. When he did not find anything, Forrestal claimed that *they* knew he was coming and had removed the bugs, only to replace them after Souers left. Stacey speculates that Raines (Bethesda's Chief Psychiatrist) would probably have known about the Souers visit and may have sanctioned it as some kind of planned therapy. However, Forrestal's condition was such that the charade (if it was a charade) did not help him at all, and his paranoia about being watched increased. (In fact, it is highly likely that Forrestal *was* watched, because of the security risk of a sick man who knew too many State secrets).

There seems little doubt that Forrestal was kept under what probably amounted to house arrest so that he could be monitored continually, despite pleas from his brother for him to be moved to a more friendly environment where he could meet other people and where he would stand a better chance of recovery. This arrangement carried on for about two months and Forrestal was then encouraged to leave his room to walk around the floor where other patients and nurses could be found. Such a relaxation of attention was officially believed to be part of the treatment, which presumably would ultimately start the healing process. Raines is on record as stating that such patient privileges could attract the real risk of suicide by the patient, but in such cases the risk had to be taken anyway if the healing programme was to be given a chance.

While for his own safety Forrestal had three military guards around the clock, the regular midnight guard had been replaced by another who did not know Forrestal and was not alert to the type of mental illness from which he was suffering. Stacey remarks that no one really knows what happened next. It is assumed that Forrestal sent the obedient guard off on some errand to get him out of the way. He then went into a pantry area, which unfortunately did not have a shielded window. He is thought to have tied one end of his robe sash around his neck and the other end to a radiator. He then climbed out of the window and jumped. The sash did not hold his weight and he crashed to his death on to a third-floor roof from thirteen floors up.

That James V. Forrestal was suffering from a form of paranoid schizophrenia is beyond doubt. That he was virtually incarcerated in a secure hospital area in case he blurted out state secrets about UFOs, the Roswell incident and his government's contact with ETs could be stretching a point. If the legends are to be believed, the fact that he became US Secretary of State for Defense in the crucial era of Roswell and that he was reputedly MJ3 of the MJ12 Panel, reporting direct to President Truman, is very significant. Whether overwork, internal politics and rivalries, together with an allegedly pressurized personal life, were the cause of his suicide, we will never know. That it was, rather, the additional burden of the knowledge of his government's contact with ETs in his position as MJ3, together with what may have been an irreconcilable head-on attack to his religious beliefs, that drove him to his death, could be a total fiction. But we can never be absolutely sure.

DEATHS AND DISAPPEARANCES

Some readers may remember the strange deaths and disappearances of UK scientists working on sensitive projects in the 1980s. There is some evidence to suggest that many of the deceased personnel were working on secret Star Wars projects for the US at the time of their deaths. Disquietingly, the Pentagon sought explanations for the deaths in the face of an apparent inability or unwillingness on the part of the British Ministry of Defence to view the situation as abnormal.

The MOD had previously made a statement saying that the incidence of suicides in defence workers is generally less than in the population as a whole. They agreed with the investigating police that there was no link between the deaths. The official line deplored the demise of so many scientists, but could not see any connection between the deaths of a number of workers on the same projects; at the same time, there was a distinct refusal to admit that twenty-two incidents was well above what could be considered a normal demographic rate of suicide for the whole population – remarkable!

The refusal to find linkages was even voiced by a GEC Marconi director, who went public to state that the company was convinced that there was no connection between any of the deaths. Even a senior MOD source said that most of the unfortunate scientists had never touched anything classified in their lives and that to his knowledge there was no connection to the US Star Wars programme.

In early October 1985, an American source had confirmed that the MOD statement was nonsense. While the source could not say how many or name names, it was clear that there *was* a Star Wars connection in some of the cases. The MOD countered by stating that if there was, it involved only non-restricted computer simulations which could be carried out at a number of universities; however, even that was unlikely.

It transpires that the US Embassy monitored these strange deaths and there is some reason to believe that the numbers probably exceeded twenty-two in total. As far as is known, the rash of suicides has now ceased, but the following list makes interesting reading, particularly as the verdicts on some of the deaths still remain open. (More information is given in an article by Gordon Creighton in *Flying Saucer Review*, Vol 36, No 2, and in *Open Verdict – An Account of 25 Mysterious Deaths in the Defence Industry* by Tony Collins – see Bibliography.)

Alphabetical list of casualties
Baker, Michael, twenty-two years old. Digital communications expert working on a defence project at Plessey; part-time member of Signals Corps SAS. Fatal accident on 3 May 1987, when his car crashed through a barrier near Poole in Dorset. Coroner's verdict: Misadventure.

Beckham, Alistair, fifty years old. Software engineer with Plessey Defence Systems. Found in August 1988 electrocuted in his garden shed with mains wires connected to his body. Coroner's verdict: Open.

Bowden, Professor Keith, forty-six years old. Computer programmer and scientist at Essex University engaged on work for Marconi, who was hailed as an expert on super computers and computer-controlled aircraft. Fatal car crash in March 1982, when his vehicle went out of control across a dual carriageway and plunged on to a disused railway line. Police maintained he had been drinking but family and friends all denied the allegation. Coroner's verdict: Accident.

Brittan, Dr John, fifty-two years old. Scientist formerly engaged in top secret work at the Royal College of Military Science at Shrivenham, Oxfordshire, and later deployed in a research department at the MOD. Death by carbon monoxide poisoning on 12 January 1987 in his own garage, shortly after returning from a trip to the US in connection with his work. Coroner's verdict: Accident.

Dajibhai, Vimal, twenty-four years old. Computer software engineer with Marconi, responsible for testing computer control systems of Tigerfish and Stingray torpedoes at Marconi

Underwater Systems at Croxley Green, Hertfordshire. Death by 74m (240ft) fall from Clifton Suspension Bridge, Bristol, on 4 August 1986. Police report on the body mentioned a needle-sized puncture wound on the left buttock, but this was later dismissed as being a result of the fall. Dajibhai had been looking forward to starting a new job in the City of London and friends had confirmed that there was no reason for him to commit suicide. At the time of his death he was in the last week of his work with Marconi. Coroner's verdict (after two post mortems): Open.

Ferry, Peter, sixty years old. Retired Army Brigadier and an Assistant Marketing Director with Marconi. He was said to be depressed following several serious car accidents, the last on 2 August 1988. Found on 22 or 23 August 1988 electrocuted in his company flat with electric leads in his mouth. Coroner's verdict: Open.

Gida, Avtar Singh, twenty-seven years old. Belonged to the MOD Admiralty Research Establishment. Disappeared mysteriously in January 1987 while writing his doctoral thesis on underwater signal processing at Loughborough University. Both mainland police and Interpol launched searches for him in several countries, without success. He eventually reappeared four months later. He had been traced to a red light district of Paris and confirmed that he did not know precisely how he had got there. Allegedly, he has returned to his work and has said that he does not want to discuss his disappearance or the death of his colleague, Vimal Dajibhai.

Godley, Lt-Colonel Anthony, forty-nine years old. Head of the Work Study Unit at the Royal College of Military Science. Disappeared mysteriously in April 1983 without explanation. His father had bequeathed in excess of £79,000, to be collected by 1987. The money was never claimed. Presumed dead.

Gooding, Stuart, twenty-three years old. Postgraduate research student at the Royal College of Military Science. Fatal car crash on 10 April 1987 while on holiday in Cyprus. The death occurred at the same time that College personnel were carrying out exercises on Cyprus. Coroner's verdict: Accident.

Greenhalgh, David, forty-six years old. NATO Defence Contracts Manager with ICL, who were working on the same defence project as David Sands (see opposite). Mysterious 12m (40ft) leap from a bridge at Maidenhead, Berkshire, on 10 April 1987 – the same day as Stuart Gooding's fatal car crash. He survived the fall and confirmed that he had no idea of how he had leapt from the bridge.

Hall, Andrew, thirty-three years old. Engineering Manager with British Aerospace. Carbon monoxide poisoning in a car with a hosepipe connected to the exhaust, in September 1988. Coroner's verdict: Suicide.

Hill, Roger, forty-nine years old. Radar designer and draughtsman with Marconi. Died in March 1985 by shotgun blast at home. Coroner's verdict: Suicide.

Jennings, Frank, sixty years old. Electronic Weapons Engineer with Plessey. Found dead from a heart attack in June 1987. No inquest.

Knight, Trevor, fifty-two years old. Computer engineer with Marconi Space and Defence Systems in Stanmore, Middlesex. Found dead at his home in Harpenden, Hertfordshire, on 25 March 1988 at the wheel of his car with a hosepipe connected to the exhaust. A St Alban's coroner said that Knight's woman friend, Miss Narmada Thanki (who also worked with him at Marconi) had found three suicide notes left by him which made clear his intentions. Miss Thanki had mentioned that Knight disliked his work but she did not detect any depression that would have driven him to suicide. However, she confirmed that he had suffered from migraine over a number of years and had also been involved in several road traffic accidents. Coroner's verdict: Suicide.

Kountis, George, age unknown. Systems Analyst at Bristol Polytechnic. Drowned in April 1987– the same day as Shani Warren (see page 154) – as the result of a car accident, his upturned car being found in the River Mersey, Liverpool. Coroner's verdict: Misadventure. (Kountis' sister called for a fresh inquest as she thought that 'things didn't add up'.)

Moore, Victor, forty-six years old. Design Engineer with Marconi Space and Defence Systems. Died from an overdose in February 1987. Coroner's verdict: Suicide.

Peapell, Peter, forty-six years old. Scientist at the Royal College of Military Science. He had been working on testing titanium for its resistance to explosives and the use of computer analysis of signals from metals. Found dead on 22 February 1987, allegedly of carbon monoxide poisoning, in his Oxfordshire garage. The circumstances of his death raised some elements of doubt. His wife had found him on his back with his head parallel to the rear car bumper and his mouth in line with the exhaust pipe, with the car engine running. Police were apparently baffled as to how he could have manoeuvred into the position in which he was found. It was confirmed that Peapell had shown no signs of stress which could have caused him to commit suicide. His death followed the somewhat similar death of John Brittan (see page 150). At the time of his death, Peapell no longer worked at the Royal College of Military Science and had moved to a research department of the MOD. Interestingly, both Peapell and Brittan had both worked at the Royal College of Military Science and, furthermore, both had been on a recent trip to the US in connection with their work. Coroner's verdict: Open.

Pugh, Richard, thirty-seven years old. MOD computer consultant and digital communications expert. Found dead in his flat in January 1987 with his feet bound and a plastic bag over his head. Rope was tied around his body, coiling four times around his neck. Coroner's verdict: Accident.

Sands, David, thirty-seven years old. Senior scientist working for Easams of Camberley, Surrey, a sister company to Marconi. Dr John Brittan (see page 150) had also worked at Camberley. Fatal car crash on 30 March 1987, when he allegedly made a sudden U-turn on a dual carriageway while on his way to work, crashing at high speed into a disused cafeteria. He was found still wearing his seat belt and it was discovered that the car had been carrying additional petrol cans. None of the 'normal' reasons for a possible suicide could be found. Coroner's verdict: Open.

Sharif, Arshad, twenty-six years old. Reported to have been working on systems for the detection of submarines by satellite. Died in October 1986 as a result of placing a ligature around his neck, tying the other end to a tree and then driving off in his car with the accelerator pedal jammed down. His unusual death was complicated by several issues: Sharif lived near Vimal Dajibhai (see page 150) in Stanmore, Middlesex, he committed suicide in Bristol and, inexplicably, had spent the last night of his life in a rooming house. He had paid for his accommodation in cash and was seen to have a bundle of high-denomination banknotes in his possession. While the police were told of the banknotes, no mention was made of them at the inquest and they were never found. In addition, most of the other guests in the rooming house worked at British Aerospace – prior to working for Marconi, Sharif had also worked at British Aerospace on guided weapons technology. Coroner's verdict: Suicide.

Skeels, David, forty-three years old. Engineer with Marconi. Found dead in his car February 1987 with a hosepipe connected to the exhaust. Coroner's verdict: Open.

Smith, Russell, twenty-three years old. Laboratory technician with the Atomic Energy Research Establishment at Harwell, Essex. Died in January 1988 as a result of a cliff fall at Boscastle in Cornwall. Coroner's verdict: Suicide.

Warren, Shani, twenty-six years old. Personal assistant in a company called Micro Scope, which was taken over by GEC Marconi less than four weeks after her death. Found drowned in 45cm (18in) of water not far from the site of David Greenhalgh's death fall (see page 152). Warren died exactly one week after the death of Stuart Gooding and serious injury to Greenhalgh on 10 April 1987. She was found gagged with a

noose around her neck. Her feet were also bound and her hands
tied behind her back. Coroner's verdict: Open. (It was said that
Warren had gagged herself, tied her feet with rope, then tied
her hands behind her back and hobbled to the lake on stiletto
heels to drown herself.)

Wash, Jonathan, twenty-nine years old. Digital communications expert who had worked at GEC and at British Telecom's
secret research centre at Martlesham Heath, Suffolk. Died on
19 November 1985 as a result of falling from a hotel room in
Abidjan, West Africa, while working for British Telecom. He
had expressed fears that his life was in danger. Coroner's verdict: Open.

Wisner, Mark, twenty-four years old. Software engineer at the
MOD. Found dead on 24 April 1987 in a house shared with two
colleagues. He was found with a plastic sack around his head
and several feet of cling film around his face. The method of
death was almost identical to that of Richard Pugh some three
months earlier (see page 153). Coroner's verdict: Accident.

There have been other mysterious deaths in the scientific community
involving bizarre and inventive methods of self-destruction which
cannot be considered the *modus operandi* of even a disturbed normally rational mind. The deaths clearly have some linkage, inasmuch
as many of them involve British scientists working on secret projects
for their own government or other scientists contracted to the UK
defence industry by the US. The number of open verdicts is also a
cause for great concern and it would be interesting to know if the
police files are still open in those cases. Many of the deaths are suspicious to say the least, and the passage of time does not contribute to
their resolution.

Whether or not stress levels at work played an important part in
these tragedies is generally not known or admitted. Furthermore,
without any firm proof it would be irresponsible to suggest that such
stress might have been caused by secret knowledge of alien technology in connection with the Star Wars programme *that should not have
been possessed.* However, that the deaths did occur in the manner
described is fact. That there is some kind of link between them is
probable, at least in some of the cases. It should be remembered that

the kind of close-knit scientific community which would be undertaking such specialized work would necessarily be quite small. In that regard, some social interaction in the lifetimes of the deceased should not be too hard to find.

The interest of the Pentagon in many of the deaths gives us cause to suspect that there is much more to the incidents than our own government has told us by way of ministerial and executive statements, which routinely dismissed the deaths as the product of unhappy coincidences that have *no link whatsoever* with one another. We are simultaneously asked to believe that defence workers are *less* likely to commit suicide than workers in any other profession or job. If that is the case, no doubt psychiatrists will have something of a field day when they attempt to describe a newly discovered type of paranoia called the 'Lemming Syndrome'. It is presumed that many of the relatives of the deceased do not share that hypothesis and would prefer to know the *real* truth behind the tragic deaths of their loved ones, especially where the verdict on their family member remains 'open'.

INSTITUTIONAL FEAR

When a soldier's resistance to fear has been lowered by sickness or by a wound the balance has been tilted against him and his control is in jeopardy, at any rate for a time. The wounded soldier has just visualized danger in a new and very personal way.

Lord Moran, writing on World War I

Humankind has had to deal with fear since time immemorial. Some would say it has something to do with the human being's indomitable spirit. The fear might be the collective fear of the genocide of your race in the face of a ruthless and implacable enemy, such as that experienced by the Jewish nation in World War II. It might be the plight of the native American Indian in the face of a technically advanced civilization, or of the Aborigine of Australia who is forced to integrate into a Western society. It might be the fear of the Incas of Peru, who nevertheless stuck to their immutable beliefs based on superstition and the power of nature, against the might of the Spanish Empire. The list is endless and always involves a race struggle between factions that on the one hand do not seem to want 'civilization' and its trappings pressed upon them, and on the other those who see financial profits to be made from a vanquished nation.

W A R

Collective fear is a fear which can be marshalled and rallied against. Wars are fought because of it. Even so, our histories show that (if no one else intervenes) victory over the fearful nation is the usual outcome. Whether the conquest results in the occupation of territory, the confiscation of property and the subjugation of people, or a gradual economic takeover without force of arms or bloodshed, the end result is the same. In the former case, the bloody war becomes assigned to a glorious history by the victors, while in the latter, no plaques commemorate a victory. Nevertheless, both types of takeover are essentially violent acts against the nation state and the person, and both write history. Genocidal conflicts are, of necessity, the more bloody and protracted, until the protagonists are eventually halted by other nations. They are the most vicious and may never be forgiven by those on the losing side, thereby sowing the seeds of never-ending war. We hold our own nations, cultures and individual ways of life very dear indeed.

It is an interesting fact that non-warring nations usually justify their non-intervention in neighbouring conflicts merely on the basis that their self-interest is not threatened or a treaty has not been broken. The natural way should be to warn off an adversary before he takes to arms. The differences between humankind and animals ensure that once a person or people has made the decision for aggression it is usually a fierce affair until the opponent is either defeated or capitulates, while the animal appears to possess a more reliable method for the survival of its species by backing off once teeth or claws have been shown. We therefore rely heavily on our politicians and mediators for survival. When they fail us, we are committed to collective social aggression, even though individually we may still disapprove strongly.

Non-intervention policies are not difficult to justify. Nations are reluctant to go to war with nations who have superior firepower or manpower. However, as the Russian Army proved in several campaigns against the advancing German troops in World War II, it is possible to fight a better-equipped opponent and still win, despite your own appalling losses. It was not uncommon for German soldiers to see an advancing horde from the Russian side charging towards them with no weapons at all. When a Russian soldier carrying his PPSh41 machine gun fell, another picked it up and continued the charge on the German lines, until Russian bodies piled up in a huge blood-soaked wall – and still the Russian infantry clambered over their comrades' broken bodies to carry on their charge. It was said that the German machine-gun barrels glowed red in their haste to mow down the charging Russian soldiers. Behind Russian lines the Commissars could be seen riding up and down the lines of troops on their white horses, forcing the soldiers to charge or die where they stood – better to be killed by the enemy than by your own.

Heroic acts of selflessness, bravery in the face of terrible odds and the certain knowledge that you probably would not survive is the rule rather than the exception at such times. Reputations are made and legends born. Loved ones, friends and comrades-in-arms are found and then lost. Friendships that last through life are forged from unlikely chance meetings, cemented by the mortar of a common cause against an enemy. Fear is never really overcome or beaten: it is merely suspended for a while as you get on with what you see as your job. After you have done your job, killed your first human being, the fear returns with a greater vengeance than before, for you are not the same person. War dehumanizes and, at the same time, transforms human awareness so that you become more human than you ever thought possible as a result of the experiences you have lived through. That is the reality of total war – the tangible and bitter taste of fear which can never be forgotten.

LIFE FEARS

While total war is perhaps the extreme of fear, there are also other kinds. We live with fear constantly in our lives. Some of us experience it daily: it could be the fear of being mugged by assailants, who may exist more in the imagination than in reality; the fear of a school examination and the expectation of failure; the fear of a surgical operation and its outcome; or perhaps the fear that you may say the wrong thing to someone you love dearly. Fear has many faces. It prevents us from doing stupid things – like stepping up to the edge of a cliff to look over. In that way, fear continually protects us from ourselves, providing a danger warning that we should heed if we are to stay healthy in body and mind. Fear also brings the exhilaration of surviving intact against the odds.

Anxiety fears can do you damage and cause you to become very ill or even to die. Such fears can be the product of irrational beliefs that you can, in fact, conquer things affecting your life over which you actually have no control – clearly, it is not logical to believe that you may be able to change things which are outside your control. Illness starts as an irrationalitiy in the mind and can then manifest itself physically as a medical condition, which is often life threatening – the so-called 'stress diseases'. We have to learn to live with the frustrations of ordinary life.

However, the demands placed upon us as individuals by our modern pressure-cooker lives give the impression of an ever-increasing upward spiral. Some people appear to escape the common fears that this induces by isolating and losing themselves in either social or religious communities. No one *escapes* fear, of course – all they do is exchange one kind of fear for another. Some may even develop other irrationalities which will in turn cultivate additional fears. Fear appears to be endemic in human beings and is used by nature as a method of restraining and controlling our more primitive irrational urges.

The Luddites of the nineteenth century knew fear: they were rightly afraid that new technology would take away their outmoded jobs. They saw change as a destructive element that would wreck their way of life and fracture their communities and culture. In the long run they were probably right, but no one can stand in the way of change, especially when that change is linked to a commercial objective.

RELIGION

The fear of eternal damnation has, up to recent times, been the stock in trade of organized religions, safe in the knowledge that disproof of theologies and interpretation of sacred texts is hardly likely to occur. Fire and brimstone threats kept the uneducated masses in check, formed power bases and served to consolidate commercial ambitions. That fear created a 'faith industry', which continues to flourish even in the face of well-established religions and seems a particularly nasty kind of decadence. In many ways, the modern Luddite is not an anti-techno freak but rather a faith-based reactionary who refuses to update or renounce outdated or outmoded thinking that, by definition, must act as a brake on modern society.

Religion seems to be undergoing something of a revival. Much of its current success can probably be attributed to the failure of modern society to provide any meaning in life, rather than to the success of any particular faith campaign – after all, aren't the messages always the same? They have never been updated – *can* they be updated? Arguably, the inappropriateness of all religion to modern life is becoming heightened as we hurtle towards the millenium. Prophesies that are never fulfilled; pronouncements that embarrass everybody when groups of believers are left sitting on mountain tops waiting for the end of the world; mass suicides in cult groups who believe the strange teachings of their mentors; rumours of the existence of powerful transmogrifying supernatural arks, lances or other hardware all serve to titillate the curious or control the faithful.

In that respect, it is not unreasonable to contend that religion appears to be a defence against intelligence. People are asked regularly to suspend belief in common sense and adopt a faith which has its roots more in spectacle, circus and magic than in science. However, having said all that, there does appear to be some kind of pattern or design to our Universe at the very lowest level which theoretical physicists find hard to explain as merely a series of happy coincidences that just happen to have resulted in life as we know it. Nevertheless, it seems doubtful that the design is more than the sum of its causes.

The spectacle of hard-up pensioners giving money they cannot afford to churches that already possess fabulous wealth; orthodox priests who run tap water into an urn to sell it as 'blessed holy water'; monstrous edifices dripping with gold into which the laity are allowed in order to see 'where God lives' – it all seems almost surreal. It is worrying that most of us seem to need to have faith in a belief which makes the supplicant so insignificant and unimportant. To drain the fire and shackle the spirit of human beings so as to reduce them to objects who are no better than obedient slaves, doomed continually to worship an unseen yet omnipotent being of which they know nothing, cannot be the mark of something which can endure forever. This kind of oppression does not allow humankind to grow. Will the message of our visitors help us to overcome our mind block?

THE ALIENS ARE HERE

Fear is rooted in societal infrastructures all around the world. We operate on fear: fear of being caught in the commission of a crime, of being discovered in a lie... what kind of fear, therefore, would the world feel if it were known that aliens are among us? Would it be a *new* fear, a fear we had not felt before? Or just a fear of the unknown, perhaps? Or maybe the fear of a takeover, an invasion? What would happen to our societies? The Western world could handle it – couldn't it? What about the Third World – no real need to tell them, is there? After all, it doesn't really affect them, does it? They will not benefit from this contact. (Of course, the Third World would not agree, as they would be the ones most likely benefit most from any technological handouts. The West surely must not get the lion's share as they have always done in the past – equal shares for all, please.)

Let's examine what an overt alien presence on Earth might do.

1 Organized religions could embrace the aliens as being just another product of a creative pulse from an omnipotent God – at least, until the aliens tell us more about the cosmos in which we live. The religionists would then be in an unsustainable defensive mode which would have very serious potential for the collapse of organized religion. There would no doubt remain splinter groups who would stick avidly to old faiths for comfort. The question is, would the aliens be able (or want) to fill the void left by the collapse of so many faiths in the world? Unlikely, for existing belief structures are the products of a nation's cultural development and it is therefore improbable that differing cultures will embrace a uniform religion where sharing has not already occurred. The vacuum may not be filled, potentially resulting in very dangerous consequences for world cultures.

2 There would be a massive human expectation of help from the aliens, once it was clear that they meant us no *physical* harm (irreparable damage would already have been caused to our societies and belief systems). However, the fact that the 'no physical harm' scenario depends largely on goodwill from the visitors, as *they* clearly possess technology which outstrips our current science, is not a great cause for confidence. If the visitors turn aggressive we could probably destroy some of them (those who remain on or close to Earth) with nuclear weapons, but this offers no comfort because in so doing we would destroy ourselves and make our planet so radioactive that those who survived could not live here anyway. The last-resort use of nuclear weapons would need to be very localized and employed in a surprise strike, if used at all. The question of the scale and method of retaliation would, of course, be considered in great detail before any nuclear buttons are pressed. However, if the nuclear buttons are not pressed we could be defenceless.

3   The balance of East and West has already been mentioned several times. It is crucial that if and when the matter of contact with aliens comes out into the open, equal opportunities should exist (assuming that *they* have no preferred option). If the UN were not so fractured and damaged, it could provide the kind of international forum required. Sadly, it has no power of its own to force any issues and can only function via the nations it represents. One possible method of maintaining balance would be for individual nations to cease to exist. In the light of past history, however, this expectation seems very naïve: just one world is a dream dreamt by many in the past, but it may *have* to become a reality if *Homo sapiens* is to survive.

4   Unless the aliens show all the world's people the same consideration, rivalries will form to split existing alliances and agreements. Hostilities will break out in savage retribution for unfair treatment and will be directed at the nation which is perceived to have more than the others. 'More' here means prosperity, increased health and life expectancy based on alien technology and an understanding of nature. If hostilities become too severe, it is possible that the alien presence would withdraw until matters are settled. It seems illogical to imagine that any alien influence would interfere overtly in human politics.

5   If the alien message were one we collectively did not want to hear – such as 'the Earth will be soon be devastated by cometary impact' – the outcome is not difficult to imagine. Chaos on an unprecedented scale would ensue and human social cohesion would break down. There would be no way of dealing with such events on a global scale. An efficient failsafe mechanism would have to be introduced to prevent nations from knowing about the forthcoming disaster. In other words,

it would be allowed to take place without the aliens' intervention – even if it were possible for such an event to be averted
by their technology.

6   What if our visitors tell us that our history is not as we
thought: that we are not the product of natural evolutionary
processes and did not evolve from ape stock (which seems
quite likely, due to the incompatibility of blood types – we
actually seem closer in evolutionary terms to the pig!), or as a
result of crawling out of the ocean, or from some divine
process of creation, but are rather the product of a selected,
genetically altered sub-species of themselves or some other
species? It is not difficult to imagine the kind of impact that
knowledge would human beings and their belief systems!

Science fiction? Maybe – maybe not. Either way, it would not be wise
to let our arrogance get in the way of exploring such possibilities.

All of these issues, and more beside, have probably already been
mapped out in some secret US location. Being custodians of the truth
must be a very heavy burden. The knowledge of alien contact is
fraught with many irreconcilable human taboos and beliefs. Open
contact would cause very serious and traumatic disruption to all
nations overnight and destroy whole tranches of tradition, belief and
faith built up over millenia. There would be a real risk of the aliens
being made Gods. For their part, it is unlikely that *they* would want to
be Gods, as such a position would incur considerable responsibility.
If *they* were here before, fine-tuning our development, it may have
suited their purpose to be seen as Gods at that time. If that was not
their intention, perhaps people made *them* into Gods anyway because
they had no terms of reference other than the religious or supernatural in which describe what they saw.

The problem with an overt alien presence is that it would not unite
us. It would actually have the opposite effect, because all nations run
at different speeds – this is a very important point. While wars cause
great devastation, death and suffering, they tend to have a unifying
effect on both sides of the conflict when it is all over. If Earth history is to be played out normally, the only unifying event that would
occur is for nations to rally against the intruder – the alien presence.
That may be one major reason why a presence has not yet been dis-
closed.

CAUTION

This chapter has been concerned with fear, on both a species and an individual level. There may be thousands of people who would be quite prepared to take a trip à la *Close Encounters of the Third Kind* with our skinny grey friends. Equally, most of us would be horrified at the prospect. Sadly, the willingess of the first group might be caused by the feeling of despair with their own lives – the feeling of 'anything must be better than this' – in order to risk such an unknown enterprise without any notion of fear. For the rest of us, maybe the degree of fear is exaggerated to an inappropriate and irrational degree. Both situations fall short of the ideal, which should be a great deal of curiosity combined with extreme caution. We should continually remind ourselves that it is fear of damage or death that prevents us as individuals from falling over the cliff edge – and thence to oblivion. On a grander scale, not to heed the natural warning of fear could mean species extinction.

LOST MEMORIES

At last it seems to me I have come to understand why Man is the most fortunate of creatures and consequently worthy of all admiration and what precisely is that rank which is his lot in the universal chain of being – a rank to be envied not only by brutes but even by the stars and by minds beyond this world.

Giovanni Pico, philosopher (1463–94).

The *Chambers Dictionary* lists civilization as, 'the state of being civilized... having advanced beyond the primitive savage state; refined in interests or tastes; sophisticated, self-controlled and well spoken'. Unfortunately, even in modern times, it is unusual to find all of those qualities in one person, let alone a nation!

In his thought-provoking book *Hyperspace – A Scientific Odyssey through Parallel Universes, Time Warps, and the Tenth Dimension* (see Bibliography), Michio Kaku relates how Nikolai Kardashev, a Russian astronomer, categorized possible future civilizations into three clearly defined types:

Type I
A civilization that controls planetary energy sources and controls the weather, prevents earthquakes, mines the Earth's crust for minerals and harvests the oceans. This civilization has already completed exploration of its solar system.

Type II

A civilization which controls the power of the Sun – mining it, not just harnessing the energy output. This type of civilization will require such power to run its machines. Colonization of local star systems will be undertaken.

Type III

A civilization that controls the power of an entire galaxy. This civilization will have harnessed the power of billions of star systems. It has also mastered Einstein's equations and can manipulate space–time at will.

Such an hypothesis seems very wild, to say the least. However, as Michio Kaku points out, the sum of human knowledge grows exponentially year on year and with each new discovery or invention. If we avoid self-destruction or accidental death from other planetary bodies or astronomical mischance, it is possible that the human race could aspire to at least some of Kardashev's projections.

THEORIES – AND ATLANTIS

There are many theories as to how and why civilizations rise and fall, and it is interesting to note that knowledge is not necessarily passed on to succeeding civilizations solely through word of mouth, folk history and legend. Many times things have to be rediscovered or reinvented. It is commonly thought that civilizations fall through their inability to withstand a change of climate, catastrophic war or disease: there is a lot of evidence to suggest that the Earth has been hit by large cometary bodies or asteroids in its history and continues to be at serious risk from time to time. However, there are some theories which go even further.

Michio Kaku reminds us that Richard Muller theorizes that the Sun is actually part of a double star system. The sister star (called Nemesis or Death Star) is responsible for the suspected periodic extinctions of life on Earth. Nemesis rotates around the Sun once every twenty-six million years or somewhere thereabouts. As it passes through the Oort Cloud (a supposed comet cloud that exists beyond the orbit of Pluto) it causes a cometary avalanche, which has an extreme impact on the solar system and causes massive debris clouds that blot out the Sun, thereby destroying life.

Strangely, experimental evidence suggests that geological sediments contain untypically large amounts of an element called iridium. Since iridium is naturally found in meteors, Kaku postulates that the iridium could betray the presence of comets sent down by Nemesis. Don't worry, though: Kaku comforts us by telling us that the twin Sun (if it exists) should be about halfway to its next cycle of destruction, which he advises as taking place in around ten million years' time. He speculates that the human race will have reached Type III status by then and there will be no need to worry about Nemesis.

World orders come and go, the most poignant in recent times being the prospect of a thousand-year Reich – quite a modest ambition compared to what has gone before!

Unravelling anthropological history and collecting strange facts held a particular fascination for James Churchward. An Indian Army Colonel in the 1860s, he also travelled extensively in the Southern Pacific islands, Tibet, Burma, Egypt, Siberia, Australia, New Zealand, the US and South America. He seems to have been preoccupied with demonstrating that the legendary continent of MU played a crucial role in the evolution and formation of the world's great civilizations. The theory was that the MU civilization spread outwards from the Central Pacific area (Atlantis?) some 12,000 years ago. Today, the visible remains of the enormous land mass would be islands such as Tahiti, Samoa and the Marshall Islands. While Churchward never proved his theory, rumours and legends persist of some kind of connection between the Babylonians, the Egyptians and the legendary MU civilization.

Numerous books have been written on the legend of Atlantis and the part it may have played in the civilization of the world. Most theories agree that if the continent existed it stretched down almost to the Antarctic, passing through the 55th Parallel. Westwards, the Parallel passes through Terra Del Fuego and eastwards, beyond the southernmost tip of Africa. According to Peter Kolosimo in *Not of this World* (see Bibliography), strange islands seem to come and go in that part of the world. A group of islands called the Auroras (given that name by the discovering ship the *Aurora* in 1762), some 2,000km (1,250 miles) east of the Falklands, went unnoticed by the Spanish Hydrographic Society in Madrid until the they were seen again by a Philippino ship, the *Princess*, in 1790. In an attempt to get precise co-ordinates and measurements, the *Altrevida* surveyed the area in 1794.

The surveyors found that there were three islands very close to the same meridian, with the following co-ordinates:

1 South Island: 53deg 15min 22sec S, 47deg 57min 15sec W.

2 West Central Island: 53deg 2min 40sec S, 47deg 55min 15sec W.

3 North Island: 52deg 37min 24sec S, 47deg 43min 15sec W.

Oddly, these particular islands, and others located in the survey, were not seen again. However, in 1856 the crew of the *Helen Baird* saw the Auroras, and in 1892 they were seen for the last time by the *Gladys*. The latter's ship's log recorded that a long island was seen with two hills, giving the impression that there were, in fact, three islands. Kolosimo mentions that they could have been icebergs, but icebergs tend to move. Islands seem to come and go with perplexing frequency in this part of the world's oceans. Could these islands be the archaic tips of the mountains of a sunken Atlantis?

THE FIRST CIVILIZATIONS

There is common acceptance among scholars that *Homo sapiens* first appeared in central Africa. However, there is an opposing view and evidence to suggest that *Homo sapiens* appeared almost simultaneously on several land masses around the world (details are given in *The World of Archaeology* by Marcel Brion – see Bibliography). In the distant geological past, say over two million years ago, human beings could have migrated across land masses which were not yet separated. The great land mass of Asia may have been home to people since the Pleistocene period (almost 500,000 years ago), as remains dating from this time have been found in central Asia. Legends persist that there are great interconnecting tunnels linking vast areas of the continent, with 'Shambala' and 'Agarth' inhabited by 'people from the stars'. It is interesting to note that there is scientific fact to corroborate the introduction of a new race resembling the European paleolithic man among the indigenous mongol type in Asia. The only problem is that no one knows where the European type came from. Unfortunately, much scientific work has been hampered by the fact that the Chinese belief system venerates its ancestors with great zeal – the ancient penalty for grave robbers was to be cut into thin slices! It is to be hoped that modern archaeologists do not suffer the same fate.

The Asian land mass also encompasses India. It is commonly thought that the Indus civilization, which flourished around 2,500BC, was of the Nile and Euphrates. Mythological India tells of *vimanas* (flying chariots) and weapons which are very similar in effect to modern nuclear and chemical weapons. The mythology claims that devastating weapons were used in wars fought on Indian soil. Whether these 'events' are just part of Indian mythology or actually took place is unknown, as there is no hard archaeological evidence to back up the claims. Curiously, there are strange parallels between some glyphs or characters made on Indus sculptures and those found on Easter Island, and it is significant that nowhere else in the world can the same writing be found. The 'Polynesian' Easter Island also connects very cryptically to many other ancient cultures all over the world.

It is by no means certain that all the occupants of the Indus valley were of the same race. The Gods that were worshipped were not common to all and cave paintings in the Mahadeo mountains bear striking similarities to similar paintings or etchings in America, Australia and Africa. Research by Indian scholars indicates that contemporaneous civilizations experiencing the palaeolithic, mesolithic and neolithic stages of evolution existed in India, Europe and Africa.

THE AMERICAN PEOPLES

It is generally agreed that the origin of the American peoples is a mystery. Cave drawings of prehistoric saurians (lizards) have been found in the North American province of Yucatan, which strongly implies that human beings were around at the same time that the saurians roamed the Earth. Human footprints at Carson in Nevada and Lake Managua in Nicaragua indicate that people were concomitant with the dinosaurs, which is in direct conflict with mainstream palaeontology. Aestheticians (mainly art historians) make a connection between the archaic arts of Asia and America, which indicates that perhaps some kind of intercourse existed at a time when the land masses had not yet separated due to tectonic forces. It is possible that the tectonic shift of continents caused the destruction of much of the life that existed in those times. Is it therefore possible that our dating of *Homo sapiens'* emergence and origins are in error?

The many theories which relate to the origins of native American peoples are usually described under three separate headings:

1   Polygenetic theory
There were several independent centres of civilization.

2   Monogenetic theory
There was only one centre of civilization.

3   Ologenetic theory
A mix of the other two theories.

A man named Elliott Smith propounded a further theory – the Heliolithic theory – which was based on the similarity of American cultures to those of ancient Egypt – it is interesting to recall that the Apache nation worshipped Amon Ra, the Egyptian Sun God. Some scholars have proposed that the Americans were the actual descendants of Adam and that the American language derives from that spoken in the terrestrial Paradise. Others propose that Americans were the descendants of Noah; it has been said that at the time of the Flood, Brazil was occupied by Ophir and Peru by Jobal. Other interpretations propose that the Americans were descendants of the Land of Canaan, who were driven out of the Promised Land by the Jews to take refuge in America.

Other theorists maintain that America is the Arsareth of the Bible, which became the land of the ten lost tribes. Supporters of these theories suggest that the ancestors of the American peoples were Carians, Hittites, Etruscans, Hyksos or Philistines. The Elliott Smith theory won recognition as the 'Manchester School' and relied on the 'Megalithic' theory that there were certain resemblances in ceramic art, pyramids, solar worship and race customs.

Stranger, and somewhat more tenuous, theories attempt to connect the American peoples with Germany, Scandinavia, Friesland and, oddly, Wales (it is said that the amazing sea voyages of Madoc ab Owen Gwynedd owe more to the belief that the Eskimos were Magdalenian men). Two other theories concerning the origin of the American peoples propose that their antecedants come from Ethiopia or were negroes from Equatorial Africa, or that they may have come from Polynesian or Malayan origins (the Oceanic theory).

Modern opinion seems to rest on the precept that the American peoples have Asiatic origins, the only argument being whether these derive from the Mongol, Tartar, Chinese, Scythian, Sumerian, Indian or Pre-Ayryan civilizations, or perhaps from the Cham of Indo-China. In 1927, Elliott Smith supported the argument that the original *Homo sapiens* was Egyptian. However, it appears that in 1937 a gentleman called Maximo Soto Hall countered by saying that the ancient Egyptians did not colonize America but were descendants of the Maya, who emigrated to Africa at some forgotten date. Some sixteenth-century cartography maintained that Atlantis formerly occupied a region between America and Europe and thus formed a negotiable bridge between the two land masses.

In 1921, an American archeologist named Ameghino maintained that *Homo sapiens* first appeared on the Earth in America in the Tertiary period. These people would have been short in stature and were the ancestors of all prehistoric people so far discovered. Ameghino called this prototype *Homunculus patagonicus*, but unfortunately did not receive much support for his claim that America was the cradle of humanity. Traces of the first Americans can be dated to over 30,000 years ago and they had already colonized the Yukon valley in the Pleistocene period. Alaska and Siberia were not dissociated from one another at that time and a wide corridor formed between the glacial mountains to the east and west of Hudson Bay and Labrador. It is proposed by some that Asiatic immigrants filtered through this route into the plains of North America. Later, the retreat of the ice would break the Behring 'bridge', thus preventing any land crossings from around 25,000BC.

Many archeologists concur that the only dates that can be *agreed* for the early American peoples derive from ornamental remains of the Maya found in Mexico and equate to about 98BC. The gigantic ruins of the peoples of Tiahuanaco in the region near Lake Titicaca are thought to be of an age lost in antiquity. Even more surprising is that there exist traces of at least two distinct peoples: the above-mentioned ruins now sunk beneath the waters of the lake, and the others standing back off the shoreline and dating back to the days when Tiahuanaco was a commercial port.

The presence of decorations depicting winged fish, which can only live in tropical waters, on monoliths situated over 4,000m (13,000ft) above sea level near Lake Titicaca is strange. It may well have been that distant geological upheavals caused this area to be altered or that the race that inhabited this land perhaps took refuge on the high plateau from something which is not immediately known.

Unfortunately, the plateau was not immune to being overwhelmed by flooding and one day perhaps succumbed to what must have been devastating torrents of water that swept in and destroyed the cities. Sculptures of prehistoric toxodons routinely appear on the ruins beside the decorative flying fish. It should be remembered that these creatures were extinct after the Tertiary period and also that the sculptures were made on monoliths which were found near Lake Titicaca, which itself has existed since the Tertiary period. All very strange indeed.

It is now known that long before Columbus 'discovered' America, both Europeans and Asiatics had landed there. It is said that very old Chinese chronicles repeatedly mention a continent called 'Fu-Sang', which some scholars believe to be America. There is also a striking similarity between the art of the Chinese, American, Mayan and Peruvian peoples. There are epic tales of the Viking 'Eric the Red' landing in America, which are substantiated by runic inscriptions on various stones and the rather odd one at Dighton Rock in Massachusetts. This peculiar stone is reported to have inscriptions of Chinese characters, Hebrew, Phoenician and Druidic letters, and more beside. However, there is good reason to believe that some ancient (and not so ancient) graffiti has been added to confuse the student!

While there is some evidence of Viking landings in America, even more extraordinary is the possibility that both Irish and Welsh landings pre-dated the Viking expeditions. This is based on the folk history of the legendary voyages of the Welshman Madoc ab Owen Gwynned. However, all these landings are themselves pre-dated by a much earlier presence lost in prehistory.

THE MAYA

The Mayan race is even more of an enigma. Legend says that the Maya flourished around 10,000 years ago and that they civilized Egypt. The sacred texts of the Mayan Bible, the Popul Vuh and the *Book of Chilam Balam*, have yielded a great deal of insight into the Mayan mind but there are some who maintain that the Popul Vuh was contaminated by the Christian Bible, although it is difficult to see how this might have happened, due to the timescales. This enigmatic race built pyramids in their jungles which poses a conundrum as fascinating as that of those found in the Valley of the Kings in Egypt.

The Mayan sacred capital of Tulan seems to have disappeared from the map – hardly surprising, when one considers the jungle terrain in which the Maya lived.

Some scholars maintain that the Maya-Quiche race emanated from the Mississippi basin, and the remains found in the Mississippi valley do bear some resemblance in their construction details to those found in Mayan cities. Others believe that the Maya occurred spontaneously when the immigrant tribes arrived in central America. The Popul Vuh tells of creation and how man was made by the Gods in much in the same vein as the Hebrew Old Testament. The four ancestral clans of the Maya-Quiche are recalled: the Balam-Quiche, Bala Ayab, Mahuacutah and Iqi Balam. Strange and frightening images of demons cover their temples and buildings, until the impression is given that the buildings themselves possess some kind of life all of their own.

The Maya were preoccupied with the chronology of time and much of their sculpture betrays that obsession. Unlike other cultures, such as the Egyptians and the Chinese, the Maya exhibited a free-flowing, stylized art which was not wholly logical in form, and it has therefore always been difficult to understand their glyphs. This is particularly frustrating for scholars, because it is known that the Maya-Quiche civilization persisted for some twenty centuries and only petered out when the Spaniards arrived in the sixteenth century.

It would seem that the Mayan writings were made ostensibly to commemorate events in their calendar. As their astronomers were also priests, it is not difficult to imagine that this would have been the case. However, no one is really certain what those events may have been. The Mayan year had either 365 or 260 days, depending on whether it was a civil year or a religious one. A day was called a *kin* and 20 *kin* made one *uinal*, 18 *unials* made one *tun*, 20 *tuns* made one *katun*, 20 *katuns* made one *baktun* – equal to 144,000 days. These intervals are thought to be based on astronomical observations by the Maya. Equally, no one is quite sure what those astronomical observations might have been.

It is said that the Mayan year is an ingenious construct inasmuch as it separates the civil and religious years. While reliant on astronomical observations, particularly the solar and lunar annual cycles, it therefore retains its mythological roots. A strange attribute of the calendar was that it seemed to depend on a wholly arbitrary date which pre-dated the Mayan civilization itself. Some believe that the date selected was that which was believed by the Maya to have denoted the actual creation of the world. It was thought that the Maya had calculated all the days since that time and had named and numbered the days bordering it, as well as its position in the month and in the 260-day year, together with the cycle of the calendar.

Remarkably, the Maya had also calculated the rotation period of the planet Venus. They divided the Venusian year, numbering around 584 days, into four segments: 236 days – morning star; 90 days – upper conjunction; 250 days – evening star; and 90 days – lower conjunction. These projections bear a strong similarity to the actual periodic year of Venus. The Mayan astronomer-priests were also aware that eight solar years were equivalent to five Venusian years of 365 days and 584 days respectively. These calculations were used in their ceremonies. Mayan manuscripts denote that the astronomer-priests were prodigious in their calculations and recorded 405 solar revolutions, which equated to an error of some 11,959 days (33 years) in computation. However, it should be remembered that even with modern methodologies that error amounts to no more than nine-tenths of one day. It should also be noted that the Maya invented and used the zero almost a thousand years before it was incorporated into what we now consider to be modern history!

The Mayans were not shy about pronouncing catastrophic endings. They predicted the end of the world after some 34,000 years or 12.5 million days (start date unknown). They pronounced that the world would die through floods and that the Sun and Moon would unleash vast rivers to drown life itself while the evil black God, represented by a huge bird screeching out loud cries, would circle the skies. The Maya eventually migrated northwards to the Yucatan peninsula. They abandoned their prosperous cities over a long period although there seems to have been little reason for them to do so, especially as their new home was much less hospitable than their old. No one knows why they migrated to the Yucatan and theories abound, such as that they were escaping from disease or that their land could no longer sustain them. Other theories concern uprisings by the peasants against the priests, who no doubt wished for even greater monuments to their intellectual obsessions. Unfortunately, the migrations served to undermine the culture, which may suggest that the uprising theory is correct. It may also be significant that the peasant population remained in the cities long after the religious buildings had become unoccupied.

The Maya clearly surpassed any other known cultures both technically and aesthetically. They pre-dated the Egyptians in their funeral rites, in believing that the dead required food in the afterlife. However, the Maya could be fierce and, like the later Aztecs, would concentrate in their battles on taking as many prisoners as possible, so that they could be sacrificially slaughtered at a later date by having their wrists slashed as a blood sacrifice to the Mayan Gods. Before you recoil in horror, remember that blood sacrifices were celebrated not only by these early Americans but also by the early Hebrews and other cultures as well.

GREAT WORKS

The working of gold and precious metals was by no means the preserve of just a few tribes in the Americas. The methods of working the noble metals seem to have been widespread across the American continent among races like the Quimbayas, the Diaguites of Argentina (said to be similar in many respects to the Incas in both culture and importance) and the Caraques. The latter were so advanced in their metallurgy that they could alloy platinum, gold, silver and iridium; while platinum occurs plentifully in Colombia, it does require smelting temperatures of around 2,000°C, which begs the question as to how they achieved the required degree of heat.

These people were great pyramid builders and their constructions pre-dated the Egyptian pyramids by thousands of years. The largest Toltec pyramid is found at Cholula, which is the religious capital for the God Quetzalcoatl. Interestingly, this pyramid is also the largest in the world, even larger that the Great Cheops pyramid in Egypt. The city of Cholula would have been the subject of an extraordinary number of pilgrimages from all over Mexico by the faithful; the occupying Spaniards perceived an opportunity and built a Christian church on top of the pyramid where the original temple had once been.

The Peruvian Inca civilization was sacked and destroyed just as efficiently as the other American nations visited on by the Conquistadors in their endless pursuit of gold. However, the geography of Peru made it difficult, as some of the Inca towns were perched atop high mountains and clothed in tropical jungle greenery, which effectively made them undetectable from the ground. One such town was Machu-Picchu. Perched almost inaccessibly on a mountain peak, it was built with gigantic blocks of stone carved in such a way that it would be difficult to insert a knife blade between the stone joints. No one seems sure whether Machu-Picchu was built as a temple or a fortress, for the strength and massiveness of its construction would be equally appropriate for either.

Apart from being extremely skilled masons and architects, the Incas also possessed very advanced medical and surgical skills. Trepanning was quite common, with metal plates being skillfully fitted to the patient's skull. Broken limbs were treated and several drugs were used for analgesic or antiseptic purposes. Instruments and recognizable surgical paraphernalia have also been found. The Incas worshipped the Sun and, in accordance with the beliefs of the region, human sacrifices were always dedicated to new building works to ensure that the Gods would look favourably upon them. Human skulls would be placed in the foundations of the building, presumably so that anyone entering it would have to cross over them.

There are scholars who believe that some of the Inca ruins actually precede the Incas themselves. However, it is known that the Incas probably used many differing styles of construction at the same time. Nevertheless, it is still interesting to consider that another, perhaps far greater civilization was responsible for the construction of the Inca cities using gigantic blocks of stone. Scholars indicate that the Incas passed through at least three stages of civilization: a primitive period up to around 200AD, an archaic period from around 200AD to 800AD, and what is called a pre-Inca period from around 900AD to 1150AD. The establishment of the Inca civilization would have taken place from around 1150AD until the Conquistadors finally destroyed them.

While retaining a Peruvian connection, Bolivia also has its share of mysterious cultures. The Tiahuanaco ruins are strangely decorated with symbolic and imaginative design. The colossal scope of the ruins, with huge stones and archways hewn in enormous rocks, remains an enigma. These ruins are equalled only by those near Lake Titicaca which were left by the Aymara civilization. It is thought that the Tiahuanaco ruins pre-date the Incas by at least 1,500 years and they include menhir and dolmen art peculiarly similar to that of Brittany and Ireland. In conjunction with the fascinating mystery of how Egyptian masons built their edifices, it is difficult to see how the early Peruvians could have handled blocks of stone weighing tonnes, put them in their positions and then carved decorations on them using only stone tools.

The Conquistadors systematically destroyed the Incas, principally because of their frenzied lust for gold. It is fortunate, therefore, that some of their treasures escaped the Spanish smelters to show that they surpassed even Egyptian examples of the art of jewellery. The Inca and other early American cultures still retain much of their mystery today, even in the face of all that we can muster in the way of modern technological processes to help us understand these enigmatic and secretive cultures.

Did they have help?

This excursion into history and the search for a first civilization has been necessary in order to discover whether or not there is any justification in believing that we have been helped along the way by visitors from the stars. There seems to be no evidence to suggest that these early civilizations, possibly going back to the time before the continents split and grouped into new alignments, had any material help beyond that of their own indefatigible will. It is quite wrong to denigrate humankind by suggesting that our feeble minds could only allow us to be victims of nature rather than users of natural events to our advantage. It is also wrong to imagine that succeeding generations have basically any different thoughts to those of their antecedants. The differences between ancient and modern man are not elemental and have more to do with culture, belief and the inaccessibility of powerful technology to the ancients. It is sobering to remember that the savage brutality of these American nations was matched by their remarkable building skills, unsurpassed even today.

Nevertheless, the ancients' artfulness at building immense stone structures from blocks of stone weighing several tonnes each may not be entirely their own. Some have theorized that there is a strange common thread running through ancient human civilizations that points to a 'megalithic age' which can be found all over the world, from Ireland to Easter Island. Is it too preposterous to imagine that there was once a 'golden age' and that one race was initially responsible for all these great works – a race which handed down its knowledge, art and science so that the early peoples might prosper, develop and remain, and which, to all intents and purposes, disappeared, leaving behind nothing of themselves except their legacies to their protegés?

Two opposing hypotheses exist. The first states that human beings are responsible for everything and no outside help was (or could have been) possible. The second states that there was help and that, more-over, some of the works we see today could not have been possible without help from someone. The first argument tends to disregard too many obvious problems and simply ignores facts which do not accord with preconceived scientific paradigms. The second, unfortunately, hands over too much to sensationalism without any evidence to support it. It is a fragile and unsupported argument which assumes that ancient civilizations were not very clever and, as such, could not have possibly achieved what they did without supernatural intervention. A better explanation is required.

Given that they had no modern chronometers or other instruments, the achievements of the Mayan astronomer-priests in the construction of their lunar and solar calendars and the accuracy of their calcula-tions is astounding. However, none of the achievements of these and other peoples in art or science can realistically be considered as *not* the product of human beings' incredibly fertile imaginations. The ancients' quest for knowledge may not have been driven by curiosity or economic need as it is today, but it was *necessary* because of their developing understanding of their environment, their wonder at and fear of nature, and their struggle to understand the cycle of birth and death. From these fears and wonders came metaphysical man, who possessed the intelligence to create religion, faith and ritual.

If there was a megalithic race which was responsible for the great building works and sculptures we see all over the world, is it possible that these were the original mythical Atlanteans, who perished in some cataclysmic accident aeons ago? When they perished they must have taken their machinery with them but, tantalizingly, they left their enigmatic art all over the world to give us just a glimpse of their civil-ization. They may have taught the ancient Americans how (and why) to build pyramids, and they in turn may have taught the comparative-ly modern Egyptians, who adapted and customized the pyramid for their own specific cultural purposes. The cult of human sacrifice to the powerful imagery of nature deities eventually gave way to an Egyptian hierarchical God: a man-made, tangible God-King who would effect centralized control over an emerging nation state. If that were the case, it is possible that the pre-cataclysmic root nation could have created a real Eden. What followed them may have been a cor-rupt and pale misinterpretation of their culture – now unfortunately lost forever.

Civilizations, like people, suffer birth, growth and eventual death, for that is evolution. If ETs had been around in those days, the impact on early man would have been so marked as still to be seen today, hopefully as an obviously non-human artefact which we could all recognize (if the artefact is *physical*). That would clinch it – wouldn't it? Despite what has been written or said in sensational publications or broadcasts over many years, there is no such archaic hard evidence on Earth which we can categorically call 'alien'. Perhaps when the Cydonia region of Mars is explored, or the strange constructions on our Moon are looked at in relation to Earth monuments, we may discover a link to prove alien intervention. Until that time, we can only speculate that help from non-human entities may have been given to early civilizations, thereby changing the course of human evolution through the gift of knowledge.

It is difficult to imagine the form that knowledge may have taken. The need and ability to build enormous stone structures seems now to have been lost. If the original purpose was to deify an ancient belief in human sacrifice, it does not evidence ET intervention in the affairs of humankind and may be more the product of drug crazed-cultures of unimaginable vice and cruelty or cultures dedicated to the mystery of birth and death.

Other possibilities may have included an alien 'watching and waiting' programme which, given the available evidence, may be more reasonable – that is, if you accept the archaic contact hypothesis in the first place. If ETs visited the Earth in early history they may have been a different race to the greys, who seem very focused on their particular mission as rather more recent visitors – from 1947?

THE MEDIA FACTOR AND AGENCY INTERVENTION

Suppressio veri suggestio falsi.
(Suppression of truth is suggestion of the false.)

When Orson Welles broadcast his story of an invasion from Mars, *War of the Worlds*, on radio in 1938 there was widespread panic, because people believed that the event was actually happening. The broadcast of a blow-by-blow account, with sound effects to back it up, was very convincing. The programme caused a great deal of distress to large numbers of people and many would no doubt agree that in hindsight its presentation was all too real and was a foolish stunt.

An interesting conjunction can be made in the overall timing of the broadcast with the hostilities in Europe to come the following year. As such, the broadcast would have been of immense value to behavioural psychologists in respect of more prosaic invasions of the terrestrial kind. The reaction of a spooked public had never before been witnessed on such a scale. Sure enough, one could study the stock market to watch panic at work, but that was a *different kind* of panic. The broadcast could have been a watershed which influenced future deliberations as to the nature and type of information that could be released to the American public. In that respect, the *War of the Worlds* broadcast may actually have influenced government policy or, at the very least, government thinking.

There were no more *War of the Worlds* broadcasts – it was, after all, the kind of thing you could only do once – but it sparked off an awakening in public awareness of matters concerning space exploration, 'men' from other worlds and the colonization by humans of other planets. The idea of space travel and exploration of other worlds was never far away. Fortunately for the astute film-makers, scientific knowledge of the subject was sparse and consequently there was a great deal of latitude in respect of scientific accuracy. Everybody who lived through that era remembers the kitsch 1950s movies with their poor acting, unlikely plots and special effects that betrayed the production's near-zero budget.

Post-war science fever

Following World War II, captured German V2 rockets and their advanced technology unknown to allied personnel served to fire the imaginations of scientist and laymen alike. At last the dreams could become a reality. There was a tremendous and irresistible impetus to turn these fearsome weapons to a useful peacetime purpose, which was fuelled mainly by the remarkable determination of Werner Von Braun. The Americans and the USSR saw immense propaganda mileage in harnessing captured German technology. Combining this with the less publicized captured research on nuclear fission (to which both the Americans and the Soviets had access), both the US and the USSR correctly believed that the masters of these powerful new secrets would be the new world rulers.

This understandable but naive approach caused a new war to develop – the so-called Cold War. The large geographical distances between the main areas of occupation on the respective continents, together with the marshalled support of friends and allies, served to heighten tensions. The diametrically opposed ideological stances adopted by the US and the USSR only increased the heat, and shed no light on how mutually to survive this extremely dangerous Cold War without its escalating into a hot war. The mistrust, fear and dislike of each other's cultures caused a split which could not be healed unless one side gave way – the fear of total mutual destruction does not make for friendly thoughts! The issue of who would be first in any scientific undertaking took on a new dimension beyond simple competition, friendly or otherwise. National pride was at stake, and the very survival of the respective nations could be an issue. It would become very important to prove to the rest of the world that one ideology was superior to the other by virtue of its achievements.

Such was the backdrop to the media culture following the end of hostilities in 1945. The US film industry was trying to establish a reconstruction programme itself and no doubt saw new opportunities to get the public's mind off the dreary subject of hard work by spoon-feeding them a diet of escapist therapy. You had to come out of the theatre feeling titillated, satisfied and, most of all, *good*, otherwise the film-makers had failed. On the coat tails of the epic productions came the new rash of movies – the science fiction genre. This was a niche market but an expanding one. Soon, the sci-fi movies would attract a following that would grow year on year, to be bolstered every time a new experiment was blasted into space or a rocket-powered aircraft was launched from the belly of a converted B52 bomber.

At the end of World War II the atom bomb, the German V2, the guided missile and the rocket-powered fighter had all made their debut. Norwegian airline pilots had already seen, and timed, objects believed to be Russian rockets on a flightpath following the Earth's curvature, not on a ballistic trajectory as one would expect a rocket to behave. The assumption of the time was that the Russians had transcended the limitations of the conventional short-range V2 delivery system – another blow in the Cold War. Could these have been the so-called 'ghost rockets' seen over Scandinavia following the end of the war?

In 1948, the tiny Bell X-1 flew faster than sound for the first time. Significantly, this aircraft used a German Walter rocket engine. Both the US and the USSR were keen to capitalize on captured German technology in order to advance their aeronautical and political ambitions. The British, however, did not seem to want access to that technology and saw the use of radio-controlled, rocket-powered models for supersonic research as preferable, and no doubt cheaper. The British seem to have concentrated on the development and production of the De Havilland Comet 106 Transatlantic jet airliner, which was thought to be the vanguard of the British commercial fleet of the future to beat the world in post-war civilian aviation.

Slap bang in the middle of enormous reconstruction programmes in Europe and the awakening of the US economy, plus the ever-present chill of the Cold War, came Roswell (see page 21). The timing could not have been more immaculate. It is probably a gross understatement to say that the US considered such a stroke of good fortune in getting their hands on a piece of alien technology a totally unexpected bonus – that is, if you believe the legend. That technology would surely place the US at the forefront of the post-war scientific race with the USSR – wouldn't it? The trouble with such advanced technology, however, is that you have no idea how it works, and while it is easy to figure out superficially what some independent functions might be, it could be impossible to fit that one function into the whole. For example, it has been reported that something akin to fibre-optic cabling was seen at the Roswell crash site. In 1947, fibre optics was a piece of hardware waiting to be invented and while scientists probably concluded the correct functionality of fibre-optic strands, they would have had no idea as to how they could be used or manufactured with the technologies of the time.

So, what did they have at Roswell? Nothing they could use, for sure. Even if the disc could have been repaired, which would have been very unlikely given the level of scientific knowledge of the day, they would have had what was essentially a curiosity, a toy – albeit a toy from another world. The ramifications of the find would be far-reaching and the government would want to keep the incident secret at all costs. If word got out and the USSR heard about the find, it could have been seen as an attempt by the US to create a piece of dangerous propaganda designed to confuse and increase tensions and, in the worst-case scenario, a precursor to a pre-emptive strike using nuclear weapons. The Truman administration would have wanted to keep the lid on this one at all costs: the prospect of the still-smouldering embers of World War II re-igniting as the result of impugned sensitivities was too great a risk to take.

They managed to keep the lid on Roswell during the crucial times. Today, it hardly seems relevant from a disclosure point of view, but innocent people became victims of the cover-up and they have never been vindicated or had their characters cleared. There seems little possibility of the US government ever apologizing or redressing the balance, for to do so would be a clear admission that a cover-up had occurred.

If you believe the legend, Truman then set up the famous (now infamous) MJ12 Panel, presumably to ensure the tightest possible control and that he could call on expert and reliable intelligence direct. The Panel would take charge of all the compartmentalized information sources available through the normal Secret Service channels and report direct to Truman. As time progressed events may have overtaken the President, as the demands of the ordinary affairs of office and the more fundamental needs of the nation would have taken priority. Possibly, the Panel had satisfied themselves and their President that the UFOs posed no threat to national security: this might have been based more on the non-aggressive nature of the discs than the fact that they could overfly and violate any air space in America without permission. The priority nature of the study of the discs would then have been scaled down and passed to the Air Force to keep a watching brief. There would have been no requirement to keep the MJ12 Panel intact once they had reached their conclusions about the discs being non-interventionist onlookers – even though these onlookers were from another world.

THE SPACE RACE

Presidents should be earthy men. That is, they should look to the needs of the common people and not have their eyes fixed on the stars. Perhaps Truman was such an earthy President, for the knowledge that a vehicle and bodies from beyond Earth were in US possession did not stop him from sanctioning nuclear bomb testing programmes to ensure that the USSR did not get in front in nuclear technology. This was understandable, as he knew that the USSR were carrying out their own research and it would not be long before they had the A-bomb too. It may have been at about this time that disinformation, misinformation and downright lies about the UFO issue became institutionalized and endemic in the US government. Public interest grew as animals were sent high into space, and the possibility of viable manned space flight received a boost when laboratory animals were retrieved safely from instrument-packed launches sent high into the ionosphere and near-space.

The Russians then launched Sputnik (a first), followed by other ground-breaking exploits, and then Yuri Gagarin – the USSR had beaten the US in putting a man into orbit around the Earth. The Americans licked their wounds and decided that they would put men on the Moon and achieve Werner Von Braun's lifelong dream. All the time they had the Roswell disc. By now, focus had shifted from the Roswell case and there were also Earth-based technologies to protect. The trouble was, ordinary people kept seeing strange lights in the sky and some were now saying that they had been taken by the 'people' behind these lights.

A new President, and MJ12 had long-since achieved its purpose – the administration knew all there was to know about the Roswell disc, its occupants and the technology. More importantly, they would have known the aliens' numerical strength on Earth, which was so insignificant that it could not cause a threat to world security. However, the US government still had to play down the public's fears and apprehensions about the Roswell incident, now that it had got out. Resurrecting MJ12 would not be the answer. It was decided to allow the USAF to be the tool to take the heat out of the growing hysteria and to deflect the people seeking answers: Project Blue Book was born. At the same time, covert action was taken by other government agencies to make a many-pronged attack discrediting those who promoted the ET hypothesis or who were unfortunate enough to have had more direct ET experiences. In this way, the government could conduct their own covert intelligence programmes – which included open ET contact – without 'UFO clubland' getting in the way. There would be casualties along the way, but apart from being inevitable this would also be desirable, as it would illustrate graphically the real risks of messing with the mythical 'ET thing'. In that respect, Roswell was transformed from a PR catastrophe into a successful intelligence tool which could be used for a considerable time to come.

THE BIRTH OF THE CIA

When World War II ended in 1945, it is a matter of record that President Truman disbanded the Office of Strategic Services (OSS) by signing Executive Order 9621 on 1 October that year. OSS functions were transferred intact to the State Department and the Department of War. The idea of a centralized intelligence service had been envisioned prior to the end of the war by the then head of the OSS, Major General William Donovan. Somewhat predictably, he recommended that a new service should be created that would co-ordinate the intelligence efforts of several departments and report to the President direct. Donovan suggested that the new service should have no domestic internal security or any law enforcement capability (further details are given in *The Intelligence War* by Col W. V. Kennedy, Dr D. Baker, Col R. S. Friedman and Lt Col D. Miller – see Bibliography).

Truman set up the Central Intelligence Group (CIG) in January 1946. The governing authority would be the National Intelligence

Authority (NIA) and the CIG would coordinate, but not supplant, the existing departmental intelligence. Under the provisions of the 1947 National Security Act, both the NIA and CIG were deactivated. In their place would be the National Security Council (NSC) and the Central Intelligence Agency (CIA). The CIA Charter states that it has two main responsibilities: to coordinate the various intelligence efforts of the US government, and to collect, evaluate, analyse, produce and disseminate foreign intelligence. The evolution of the CIA has ensured that it is now the primary US agency responsible for the analysis of intelligence, and clandestine (Human Intelligence – HUMINT) as well as positive covert action. Development of aerial reconnaissance systems using both aircraft and satellite systems is also an important function of the CIA.

Langley, Virginia (near Washington DC), is the home of the CIA. Other CIA facilities include the principle technical and training facility known as The Farm at Camp Perry, also in Virginia. There are, of course, many other locations of interest to the CIA which are scattered all over the US (and, with permissions, the rest of the world). The organizational chart (Fig 6) shows what the internal shape of the CIA would have been at the time of the Barney and Betty Hill abduction case (see page 15). It seems reasonable to assume that the CIA would have had some part to play in this case and it is interesting to speculate as to which departments might have been involved. Unfortunately, the listings under 'Plans' are not available. However, some of the functions of the various other departments can be gleaned from this early chart. The 'Plans' section would become the 'Operations' section at a later date, as follows:

Office of Communications
Responsible for secret communications between CIA HQ and overseas locations. Installation and maintenance of radio transmitters for their own as well as agents' use is carried out by this department.

Office of Security
Responsible for the security of CIA facilities and personnel both in the US and abroad.

Office of Training
Operates and maintains training facilities, including Camp

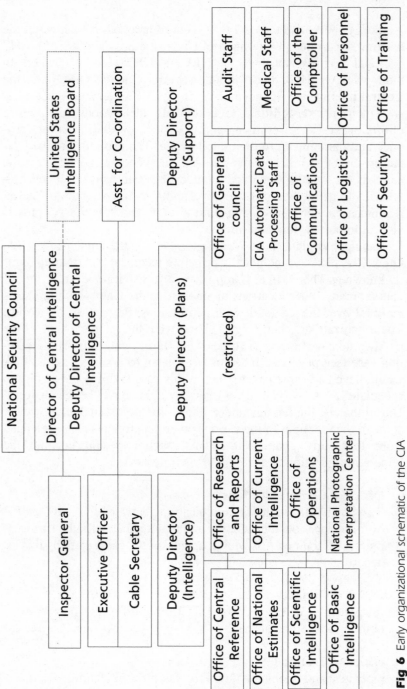

Fig 6 Early organizational schematic of the CIA

Perry. Some of the Junior Officer Trainee (JOT) programme tasks to become a CIA Case Officer were:

1   Physical training (including martial arts training).

2   Propaganda.

3   Infiltration and exfiltration.

4   Targeting and penetration of enemy, youth and student organizations.

5   Labour operations.

6   Liaison operations (with friendly intelligence operations).

7   Anti-Soviet operations.

8   Communist Party penetrations.

9   Soviet/satellite operations.

10   Paramilitary operations.

11   Agent handling (individuals and networks).

12   Foreign intelligence operations.

13   Operational and intelligence reporting.

14   Intelligence collection methodologies.

15   Counter-intelligence operations.

16   Communications.

17   Codes and cyphers.

18   Agent recruitment.

19   Meetings.

20   Security tradecraft (tools and techniques to keep operations secret plus technical skills).

Technical Services Division
This Department reported to the Deputy Director of Plans. It devised and trained personnel in the techniques and methods of:

1   Secret writing.

2   Audio operations (bugging or intercept communications devices).

3   Photography (hidden cameras).

4   'Flaps and seals' (clandestine letter-opening techniques).

5   Lock picking.

The Department also trained specialists known as TOPS (Technical Operating Specialists). In the 1970s the Department would be relocated to the Directorate of Science and Technology. The title of Deputy Director of Plans would also be changed to Deputy Director Operations.

Based on what is known of the CIA at the time of the Hill case, it would appear that the organization was poorly equipped to investigate (or suppress) such matters. However, there is little doubt that this particular US agency would soon come up to speed.

The somewhat more modern CIA Organisational breakdown is shown in Fig 7, from which it can be seen that the organization had now become much more focused in its approach. The importance of science and technology is illustrated by the straight line reporting to the Director on this chart, compared to the route to the Deputy Director (Intelligence) with no straight line to the Director of Central Intelligence in Fig 6. The Directorate of Intelligence and the Directorate of Science and Technology exist to perform the true CIA intelligence functions.

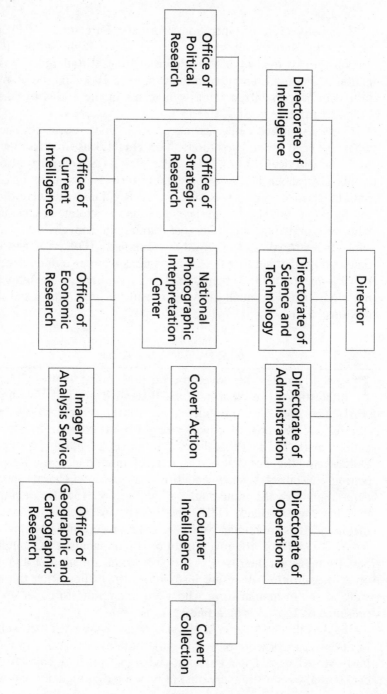

Fig. 7 Internal organization of the US Central Intelligence Agency, from a US Senate Select Committee Report

The Directorate of Operations (formerly Directorate of Plans) is perhaps the most well known. Occupying a high profile, it was responsible for the Bay of Pigs fiasco in 1961, the attempted assassination of Patrice Lumumba in the Congo in 1960 and the co-operation with US organized crime syndicates in the 1960s to take out Fidel Castro, President of Cuba.

Countless other embarrassing episodes have occurred, such as when the head of the Directorate, Max Hugel, was fired for 'unusual business practices'. Later, Alexander Haig, US Secretary of State, would accuse the USSR of organizing terrorist activities in Libya, only to discover that it was not just the Russians who were orchestrating events but also former agents of the Directorate of Operations, who were assisting, supplying *and* training the Libyans in the use of sophisticated explosives and other technology. The error was compounded by the fact that the former agents were also in contact with serving agents of the Directorate. The Directorate of Operations did not clothe itself in much glory as a result of these errors and indiscretions.

MEN IN BLACK

Fig 8 illustrates the governmental positioning of the CIA. It is interesting to see that both the Federal Bureau of Investigation (FBI) and the CIA provide input to the Director of Central Intelligence, thereby effectively straight-lining to both the President and his staff. The fact that there is no perceived liaison between the Defense Intelligence Agency, Bureau of Intelligence and Research, National Security Agency and FBI may cause problems of inter-departmental rivalries and duplication of effort. However, it could be that the 'Men in Black' (MIB) role is exclusively an FBI function – that is, if MIBs are human. While speculation continues as to whether the MIBs are (or were) government or USAF agents, UFO cult members trying to get their hands on first-hand information or a range of other possibilities, there are also some who believe that the MIBs were aliens in human form who were attempting to cover up their presence on Earth for nefarious reasons.

Wild as that may seem, MIBs nearly always visited their subjects in threes, always wore the same dark clothing and spoke in a peculiarly stilted way. Their transport (with the driver remaining in the car) was always a strangely old but new-looking model of car which

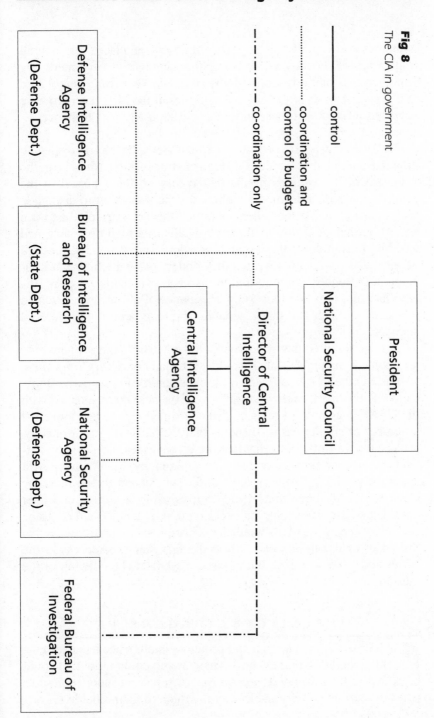

Fig 8
The CIA in government

witnesses were quoted as saying looked out of place. The overall impression of the MIBs was that a threatening and very menacing approach was being made, perhaps akin to a visit from stereotyped 1920s gangsters. In any event, the activities of the MIBs seem to have subsided in modern times since the first reported visit to Albert K. Bender in 1953.

Mr Bender was an early UFO researcher who had managed to organize a large investigations group which seems to have been in the vanguard of the various groups of the time. A letter posted in the group's magazine informed members that he was about to dismantle the organization. He told his readers that the saucer riddle had been solved but he could not divulge any details as a high-up source had told him to shut down the research. Readers were warned to proceed very carefully with any research they undertook in the future. It transpires that Bender admitted that he had been visited by three MIBs, who had told him that they knew the names of all of those involved in saucer research and that it wouldn't be long before they were all silenced.

Bender admitted that if he divulged the MIBs' identities he would go down in history but also would go to jail for a very long time. Discussing the matter, a Pentagon spokesman for Project Blue Book, Col G. P. Freeman, confirmed that the MIBs were not connected with the USAF in any way. Freeman also confirmed that he had looked at a number of similar cases. It seems that the MIBs have, fortunately, relaxed their efforts somewhat in more recent times. These early incidents may, therefore, have been an attempt by some government agency (the CIA or FBI?) to suppress the subject and at the same time cloak it in a mystique that would be irresistible to cultists at a later date. The MIBs also appear on occasion to have confiscated or stolen vital evidence. If the MIBs were aliens in human clothes, it seems that they may finally have woken up to the fact that 1930s dress in the 1950s and 1960s is weird, but in the late 1990s it is totally out of the question.

INTELLIGENCE AGENCIES

Fig 9 illustrates the known and principle intelligence agencies currently operating in the world. There are no doubt more that could be added to the list, but as the omissions represent more repressive regimes there is no reasonable way in which to gain reliable knowledge of these less organized and established national agencies. The contention by the US and UK governments that the discs pose no threat to national security is undoubtedly a smokescreen and it is unlikely therefore that any of the defence intelligence agencies would not have knowledge of the discs. The real question is whether the shared knowledge is of equal value – in other words, what is the *overall quality* of the information?

The assumption that the US knows all about the discs, their occupants and their purpose may indicate that the information given to those nations not 'in the know' would be sparse indeed. It may well be that *any* information given to even a trusted and long-standing ally would not be the *whole* truth. Realistically, that is likely to be the case, especially if the information had unpleasant connotations or, more specifically, if the data held special relevance for the US national interest. That the CIA have been and are currently involved in the UFO scene seems to be beyond question.

In *Above Top Secret* (see Bibilography) Timothy Good reveals that almost a thousand pages of data amassed by the CIA have been released under the US FOIA, with an estimated tens of thousands of documents still in CIA custody. Good also relates Todd Zechel's story of how the CIA infiltrated the National Investigations Committee on Aerial Phenomena (NICAP) in order to take over the leadership and thereby control the unruly child that was NICAP.

Zechel comments that it is perhaps not entirely coincidental that the founder of the CIA's Psychological Warfare Unit sat on the NICAP board for almost twenty years and that Charles Lombard (according to Zechel, a former CIA operative) wanted Alan Hall, a retired CIA executive, to run NICAP. Zechel theorizes that Major Donald Keyhoe was fired from his own organization in 1969 because he had turned his attention from the USAF to the CIA. Keyhoe could have been too close to discovering the CIA's involvement in covering up the increasing number of incidents.

It is reasonable to think that Keyhoe may have concluded that he had been getting nowhere with the USAF due to obstruction by the CIA. It seems that NICAP had become a thorn in the side of the US administration. The eventual outcome of the takeover by ex-CIA staffmen was predictable and NICAP seems to have been emasculated to the point where it could no longer function as an organization. In 1973 its files were handed over to the Center for UFO Studies (CUFOS).

THE MAIN INTELLIGENCE AGENCIES (Prior to USSR Reconstruction)

NAME OF AGENCY	FUNCTION	CONTROLLED BY	No. OF STAFF	BUDGET
France				
DGSE (Direction Generale de la Securite Exterieure) formerly SDECE	MI/SI/EL/FCI	Prime Minister	?	?
Deuxieme (2d) Bureau	MI	Minister of Defence	?	?
DST (Bureau for Defence and Surveillance of the Territory)	DCI	Minister of the Interior	?	?
Czechoslovakia				
STB (State Secret Security)	?	Minister of the Interior	?	?
East Germany				
SSD (Ministry for Security & Intelligence)	MI/SI/EL/DCI/FCI	Politburo	?	?
West Germany				
BND (Federal Intelligence Service)	MI/SI/EL/FCI	Chancellor	6,000	$90,000,000
MAD (Military Intelligence)	MI/EL	Minister of Defence	?	?
BfV (Counter Espionage)	DCI	Minister of the Interior	?	?
Israel				
Mossad (Office of Intelligence & Special Missions)	MI/SI EL/FCI	Prime Minister Chief of Staff Defence Forces	1,500–2,000 7,000	? ?

NAME OF AGENCY	FUNCTION	CONTROLLED BY	No OF STAFF	BUDGET
• UK				
The Secret Service (MI6)	MI/SI/FCI	Foreign Minister	?	?
Defence Intelligence Service	MI	Minister of Defence	?	?
The Security Service (MI5)	DCI	Home Office Minister	(restricted)	?
GCHQ (Government Communications HQ)	EL	Foreign Minister	10,000	?
• USA				
CIA (Central Intelligence Agency)	MI/SI/DCI/FCI	President	15,000	$1,500,000,000
DIA (Defense Intelligence Agency)	MI/SI	Secretary of Defense	7,000	$9,000,000,000 (inc NSA)
FBI (Federal Bureau of Investigation)	FCI	Attorney General	?	?
NSA (National Security Agency)	EL	Secretary of Defense	?	?
Bureau of Intelligence & Research	SI	Secretary of State	325	$12,500,000
• USSR				
KGB (Ministry of State Security)	MI/SI/EL/DCI/FCI	Politburo	25,000 (intell. out of 400,000)	?
GRU (Chief Intelligency Directorate General Staff)	MI/EL	Ministry of Defence	25,000	?

Key:
MI = Military Intelligence SI = Strategic Intelligence
EL = Electronic Intercept DCI = Domestic Counter Intelligence
FCI = Foreign Counter Intelligence

Fig 9
The main Intelligence Agencies (prior to USSR reconstruction)

The authors of *Clear Intent*, Fawcett and Greenwood, theorize that the CIA needed to infiltrate NICAP for three main reasons:

1   To gather intelligence through NICAP's investigator network.

2   To identify and plug leaks from government sources.

3   To monitor hostile intelligence agencies (KGB?).

These opinions are endorsed by Timothy Good. The authors also suggest that it is a salutary lesson that any UFO group which becomes prominent and effective may invite the same fate as NICAP.

INCREASED ACTIVITY

The 1950s seems to have been the decade of the most unusual saucer activity. It appears that the upsurge in sightings caused an invasion paranoia in several developed nations. With the sightings came the inevitable reports of crashed saucers: the 1947 Roswell incident sparked off stories of crashed discs that came to earth not just in the US but all over the world.

In 1955 a Col Gernod Darnbyl of the Norwegian Air Force stated publicly that a UFO had crashed near Spitzbergen, Norway. (This incident is, allegedly, not connected to a similar incident the same year, again near Spitzbergen, when Norwegian jets had experienced radio interference while flying near Himlopen Straits. The jet pilots are said to have discovered the cause of the radio interference in the shape of a huge blue-coloured disc lying clearly visible on the snow. This disc was around 38m (125ft) in diameter, with some kind of clear plastic dome on top. It appeared to be a remotely controlled device which may have been powered by forty or so jets around its rim. Speculation that, because of the presence of Cyrillic writing, the device was of Russian origin was thought to mean that the Russians had attempted to build a flying disc. The object was supposed to have been taken to Narvik for analysis. Needless to say, the Norwegian government declined to confirm that this event had ever taken place. It is interesting to note that the same reticence was not shown almost three years later.)

Col Darnbyl was emphatic in stating that the disc had not been made by any nation on Earth. He quoted experts who had carried out tests and who confirmed that the material the disc was made of was of unknown origin. Interestingly, he indicated that further findings would be released as soon as '...some sensational facts have been *discussed with US and British experts*. We should reveal what we have found out, as *misplaced secrecy might lead to panic*.' (my emphasis).

Unfortunately for Col Darnbyl, he had not realized that both the US and British authorities had placed a tight security clamp on the whole subject long before 1955. The promised report never materialized and presumably the good Colonel was silenced under threat of censure.

A hoax? James McDonald, an astrophysicist at the University of Arizona, is on record as stating that the reason the USAF adopted a debunking approach was because the CIA had asked them to do so in the face of the tremendous upsurge in sightings in the early 1950s. McDonald then admitted that he felt the CIA no longer operated a restraint on the flow of UFO information, but he then accused the USAF of adopting the old CIA policy in their reporting! Unfortunately, McDonald then lent his enthusiastic support to the new, and ill-fated, Condon study at the University of Colorado, which would ultimately result in the ufologically infamous Condon Report.

While it is difficult to try to understand the emotional tensions of the period from today's perspective, it is not difficult to see that professional rivalries may have occurred between McDonald, who was pro-ET hypothesis, and Condon, who represented the more traditional, objectively scientific mainstream approach. The USAF may also have been upset by the fact that they were so clearly impotent in the face of these invaders in their air space. There would have been considerable embarrassment and an element of fear attached to every incident especially those where their own pilots saw something unusual close up.

The pressure on pilots from their administrators may have been intolerable at times, but they might have taken some comfort from the fact that the other service groups would also have seen strange things on the ground and at sea. The CIA had already set up the logistics to collate, analyse and disseminate advice on policy. At the same time, they would act as the sole conduit to the President for intelligence on these matters, in accordance with their mission statement. The creation of the CIA did not immediately dispense with any of the already existing intelligence services, but it rapidly evolved as the main intelligence agency by virtue of the fact that its Director had immediate access to the President.

It is not sensible for any nation to put all its intelligence eggs in one basket and it is reasonable to think that the internal intelligence services of the US armed forces would still provide first-hand input to CIA operators. While there is no reliable evidence to corroborate the fact that intelligence on the UFO problem was shared among developed nations, it is not unreasonable to consider that some level of co-operation probably did occur. The UFO issue would then have become a shared problem, with linked intelligence networks across the developed world, MI6 and the Royal Air Force being those involved in the UK. Despite public statements that the discs posed no threat to any national security, the public still sought reassurance. Common sense suggests the official line was nonsense – and still is.

The CIA may not now have a major role to play in the UFO problem. The world has turned several times since 1947 and the subject has developed enormously. It is likely that the CIA still plays some part in the overall picture, but within the confines of its particular expertise. It may even have scaled down its operations on UFO group infiltration or handed the task over to a smaller, tactical group who operate even more covertly than the CIA.

However, evidence based on rare witness descriptions such as the Bob Lazar revelations (see page 77) seems to indicate a new twist to the story: it could be that the US government is becoming more relaxed about the discs. While it is reasonable to continue a security clamp on conventional leading-edge technology, it is not realistic to attempt to cover up the revelation that Earth is being visited by life-forms from another part of the Universe. This line of thought leads one to the possibilty that the US is perpetuating a cover-up because of the *integration of alien technology into their own programmes*. Of course, it could also mean that the keepers of secrets have become rather blasé and contemptuous of the American people, safe in the knowledge that Area 51 is an impregnable mountain fortress secure from even space satellite cameras and that there is no real risk from the saucer hobbyists. The US government may hope that the disinterested public at large will understand that there are national secrets to preserve and that the testing that is (or was) undertaken at Area 51 is just another very secret part of the defence programme to keep America ahead in the new millenium – a worthy new agenda for the secrecy weavers.

AN INSECT TRAPPED IN AMBER

Surprise, when it happens to a government, is likely to be a complicated, diffuse, bureaucratic thing. It includes neglect of responsibility, but also responsibility so poorly defined or so ambiguously delegated that action gets lost. It includes gaps in intelligence that, like a string of pearls too precious to wear, is too sensitive to give to those who need it. It includes the alarm that fails to work, but also the alarm that has gone off so often it has been disconnected. It includes the un-alert watchman, but also the one who knows he will be chewed out by his superior if he gets higher authority out of bed. It includes the contingencies that occur to no one, but also those that everyone assumes somebody else is taking care of. It includes straightforward procrastination, but also decisions protracted by internal disagreement. It includes, in addition, the inability of individual human beings to rise to the occasion until they are sure it is the occasion – which is usually too late...

Thomas C. Schelling, from the foreword to *Pearl Harbor: Warning and Decision* by Roberta Wohlstetter

Since 1947 there has been steady progression in the visitors' agenda and our response to them. Unfortunately, that agenda has been the subject of a covert and hidden series of responses made by what appears to be just one nation – the US. Whether this is because the US is more suited to that function for commercial, technological or even geographical reasons is unknown. Whether there is an historical connection between continental America and the aliens is unknown. Whether the aliens selected the US by virtue of the fact that it is the largest current energy consumer in the world is unknown. Whether it is all just the result of an accident is unknown. Whether other nations have been content to let the US speak on their behalf is also unknown. Yet all these unknowns do little to diminish the perception that the US is very heavily involved with the subject of our visitors and may be the only hosts on our planet willing or able to undertake such a liaison programme – that is, if there was a choice to begin with.

ESCALATION

It is clear even to a novice student of the subject that there has been a distinct escalation and progression in events since the 'landmark' Roswell incident (see page 21). The increasing sightings of strange aerial objects continue apace. The discs have been seen all over the world, in the air, in space, on land, on sea (and sometimes *under* it), in valleys and in mountainous areas – in short, everywhere. Many UFO sightings can legitimately be interpreted as returning space 'junk' – defunct orbital satellites or natural phenomena. Other objects could be of man-made origin, such as the perceived increase in the comparatively new stealth technology from the US. However, strange discs, pods, lights and other devices continually defy known physical explanations.

Reports of abduction are also on the increase, and while some may be examples of the much-vaunted 'false memory syndrome' (in other words, imagination), many more seem tantalizingly real, especially when there is corroborative evidence or witnesses to an event, or the appearance of body marks which defy explanation. The reputed taking of human sperm and egg samples from abductees is also on the increase, together with the terrifying new element of alien implants being secreted in the victim. Does this indicate that it is becoming more difficult for the aliens to select targets randomly, or are *they* simply becoming more confident?

This escalation is real and can probably best be described as an 'incident ladder'. It might look something like this:

1 Strange aeroforms seen in our air space. *They* seem to be testing our environment to see if *they* can function within it.

2   Tentative exploration of the flora and fauna of the Earth. Extreme shyness and caution when in contact with humans.

3   Systematic survey of Earth, showing particular interest in power origination and distribution, fresh-water supplies, military installations, weapons systems and their limitations and nuclear weapons capability.

4   First abductions. Medical-type examinations carried out on both sexes, with body fluid samples being taken, i.e. sperm and ova.

5   Discs begin to crash. SAGE (early defense radar) interference? Earth weather systems?

6   Some nations get their hands on alien hardware and pilots (dead or alive).

This is where it gets tricky: let's extrapolate on what we have so far:

7   After some crashes of discs where the occupants have survived, a deal is eventually struck between *them* and the US, using a closeted scientific arm of the US military (the US Navy?), on a technology-exchange programme or a programme which gives the greys a safe, protected haven or base(s).

8   The programme is underway and the US gets its hands on revolutionary technology for which it has no application or use. The aliens in turn have little or no need of our technology, but *they* do require some other co-operation. The number of secure bases is increased.

9   The US Navy attempts to construct its own discs but meets with failure.

10   The deal reached with the aliens is to allow *them* a reasonably free rein in using the plentiful human resource for their research into gene-splicing hybridization programmes. In return, the aliens will provide some of their non-renewable technology for the US scientists to learn from. Cattle body parts are routinely taken either by aliens for their own purpose or by US scientific groups to spread confusion and deflect attention from the main activities.

11 Initially, other Earth nations were involved in the exchange, but only the US is left in the forefront now as the aliens do not want to do business with everybody – they see it as counter-productive and a risk to their programme. There have also been some accidents which were caused by misunderstandings on both sides, several resulting in multiple human fatalities.

12 The US becomes fearful that the situation is getting out of hand but there is little they can do about it now it as the aliens are in charge at their given sites, particularly in Puerto Rico, which the US allowed *them* to have as an ideally positioned base in the North Atlantic ocean and a reduced land overfly to the Pacific ocean.

Fiction? Maybe, but there is a great deal of anecdotal evidence to suggest that some of the above may have more than a little basis in fact. Say it was all total nonsense and the US Secret Service was engineering all these crazy stories to protect a military or economic secret – is it likely that *any* country would go to such lengths to protect a *conventional science programme*? For a start, the US security clamp does not have any control whatsoever over the various disc incursions into other sovereign nations' air space. If the discs were made in the US, it is hard to imagine why they would risk such a valuable asset falling into the hands of a foreign power, perhaps even a hostile one.

It is a well-known fact that *no secret* can be kept secret forever. Intelligence services will root it out sooner or later, if left to their own devices. The Soviets will have studied their own satellite pictures of Area 51; they are also no doubt aware of the technology – conventional or otherwise – being held there through their HUMINT network. (It is understood that some Russians are currently peddling spy-satellite photographs of Area 51 allegedly showing a disc in flight!) Is it unreasonable to consider that, say, the French and British may both have some data on these secret US operations from their own satellite surveillance systems, or from intelligence exchanges with those nations who do have the information? While much of the current hype and wild claims may be laid at the door of the more extreme of the ufologists, there is no way that what is essentially a bunch of enthusiastic amateurs could construct such a web of almost believable intrigue.

The interesting rumour that the US has tried to duplicate the alien technology they allegedly possess (the Cash/Landrum affair, where a presumably crippled, highly radioactive vehicle was seen to be escorted by numerous conventional helicopters – see page 67) leads one to believe that Bob Lazar could be correct when he says that the alien technology itself is not duplicatable as the Earth does not have the heavy element 115 needed to fuel the alien reactor (see page 78). If the gravity wave reactor exists, surely it is credible that scientists would seek a solution to the problem by using readily available heavy elements, presumably uranium or plutonium? If that was the case in the Cash/Landrum affair, it clearly did not work. If all the US possesses is a few workable discs with no prospect of further fuel supplies and/or no means to duplicate the technology, the discs are no more than high-tech toys from another world.

Consider for a moment that it may be likely that the greys are the 'techno' species of alien visiting the Earth. Perhaps their specialism and responsibility among the numerous aliens supposedly visiting us is a biological and technological function? It could be that *they* are part of a greater hierarchical system in which all our visitors play a part. If that is the case, we should expect more strangeness to come.

However, it seems very likely that the greys were the first alien group to experience meaningful contact with humankind, perhaps from their own necessities. Future history may confirm that the initial contact may have happened through chance rather than design. The beneficent, kindly, gangling aliens portrayed in Steven Spielberg's *Close Encounters of the Third Kind*, who wait to give willing selected humans a celestial ride in their enormous mother ship, cultivates an implied, if not unclear, purpose, and is a distortion of what *really* happens. *Real* abductees may not have shared in the general sentimentality of the film.

BIOLOGY, GENETICS AND MIND CONTROL

A strange but nevertheless important aspect of alien contact which seems to be totally ignored among all the reports of abductions and the traditional medical examination, is the subject of bio-hazard. This is peculiar, as NASA took great precautions with the potential biological hazard in sending men to the Moon and back. Put crudely, they did not want to take or bring back unwelcome bugs. (Fortunately, it is said that NASA found that the Moon was sterile and clear of bugs, and if man did accidentally take bacteria to the Moon it would not have survived in the hostile environment.) It is very surprising, therefore, that in what is basically a meeting of two incompatible life-forms – humans and aliens – the greys appear to take such a cavalier attitude not only to our health, but also to *theirs*.

Perhaps there are clues here. Is it possible that the greys, in particular, are immune to Earth bacteria and viruses? Are we immune to theirs? Do *they* have bacteria and viruses where *they* come from? Why do abductees nearly always comment on the unpleasant aromas in the medical examination rooms? Reports of abductions do not usually include any screening or decontamination procedures. The nearest we might get to that is when some abductees remember being told to wear an unusual disposable 'paper' smock or coverall clothes while undergoing physical examination or afterwards. The parallel with conventional operating theatre clothing is marked, but it is difficult to imagine how the wearing of such garments could facilitate some kind of reliable bio-cleansing system. Could the wearing of such clothing instead be an attempt by the aliens to make the victim connect with *their* medical procedures by attempting to mimic our own?

It seems that no attention is paid to this important aspect by our abductors (or rather, the abductees in their hypnotic regressions). It would be interesting if some abductees could remember whether or not they were suffering from, say, a cold or fever at the time of abduction, or whether they contracted any colds or fevers subsequently. They certainly should, judging by all the reports of people being whisked away in the dead of night or early morning wearing just their nightclothes, and sometimes being returned to their beds cold and soaking from being around wet night-time vegetation.

Some researchers believe that the greys are somehow altering the human race for an unknown purpose, messing around with our genetic coding, experimenting on us as though we were some kind of rare and valuable livestock, growing on immature human foetuses or cloning human–alien hybrids. Some believe that *they* are predators and are predating upon us – that is, all life on Earth, not just *Homo sapiens* – for their own ends. Others believe that *they* are here on Earth to save us from ourselves now that we possess the power to destroy all life, or to undertake some kind of astral Noah's Ark to preserve the flora and fauna of our planet in the face of a natural catastrophe not yet realized.

However, could it be that *they* are giving us *their genetic coding* so that we may survive the next evolutionary step in our Earth's periodic tendency to slough off life from its face, this time through the diminishing ozone layer and the 'greenhouse effect'? If life is rare in our Universe, the motive for assisting us to survive may have little to do with their love for us and rather more with the protection of sentient life *per se*.

Is it also possible that *they possess* our planet? Perhaps *they* have always possessed it. If that is the case, we have had little determination over our lives, our history and our cultures. This depressing theory is very hard to swallow because of the diversity of human life in our world. Is it possible that we have occasionally been helped or perhaps inspired along the way by some kind of peculiar and periodic collective nervous jolt which has the tendency to lift human awareness? Perhaps it is a product of our observation of *them* and a reaction to their presence which gives the impression of their always having been here, a product of our scientific misinterpretation of space–time and our struggling understanding of matter and intradimensional relationships. If that is true, it is possible that *they* have only been here on our planet for a very short while, as far as *they* are concerned. If *they* possess the technology to shrink space–time so as to cover unimaginable distances in a fraction of our timeframe reference, it could be that *they* would already have explored most of our galaxy, if not our entire Universe, in the search for sentient life.

The most recent developments include many tales of individuals being 'tagged' by the aliens. It is presumed that the reason for implantation is to facilitate tracking of the victim at a later date. But the tag may be more sophisticated than we think. Most of these so-called tracking devices appear to be inserted high into the nasal cavity close to the frontal lobe system, but they can be planted anywhere on the body. It is possible that the devices are not what they seem and may control the victim in an unknown way. If that is a correct assumption, the whole issue takes on a more sinister dimension. Perhaps such a small device would operate within a set of closely defined parameters and instructions. It is, therefore, possible that the implant could function as:

1   Some kind of on-line system giving continual biological feedback for monitoring purposes.

2   Some kind of signal generator allowing remote tracking to be accomplished.

3   An unknown biological experiment which is removed at some later date for analysis.

4   A method of adapting or merging two disparate physiologies through chemistry.

5   A brain-control function (passive receptor device, one-way only).

The fact that no recognizable discrete components have been discovered from the rare instances of retrieval seems to indicate that signal generation may not be the purpose of these small devices (in respect of our science, of course – we would use discrete components). The on-line hypothesis is also unlikely to be correct, for the same reason. The explanation of a merging mechanism through some kind of chemical interaction is also unlikely to be correct as other, more reliable methods are no doubt available to entities who can bridge vast distances with ease and violate our Earth with total impunity. There is evidence to suggest that some implants are removed at a later date, presumably for analysis or when the device has fulfilled its function. This leaves the brain-control hypothesis. Implants are usually detected in individuals through routine medical procedures such as X-rays or as a result of a patient complaining of a local swelling or discomfort. Unfortunately, it seems that we are largely unaware of the seeding process itself, such is the perfection of the amnesic procedures.

It has been strongly rumoured that both the USSR and the US have experimented with beamed mind-control systems. The concept is concerned with controlling the mood of your enemies so that you may overcome them without difficulty or danger to yourself. The use of chemical relaxants in the form of gas or particle sprays may not be as sure as an aimed particle beam of radio frequency or microwave energy. There is no suggestion that our visitors use the same system, but the principle may be the same. The fact that some witnessess hear a buzzing sound in their heads just before a sighting occurs does suggest some kind of control mechanism at work, which presumably could be used for somewhat less prosaic purposes than an announcement of presence.

Unfortunately, it appears that recovery of these unusual alien devices is uncommon or even largely unknown due to their transient nature. They either dissolve or mysteriously disappear on the victim's demise, or they are reported to have been removed surreptitiously by the aliens at their leisure. If the reason for an implant is control, there is little real evidence to show how the control (in our normal understanding of the word) actually works. However, the subtlety of the control may be escaping us. Buzzes in the head to herald an audience with the visitors hardly seem worth the effort. Strangely, victim symptoms appear to start and finish with nosebleeds, with no particular or unusual compulsion to do anything out of the ordinary. It is worth remembering that victim compulsion had been recorded long before the current rash of implant claims. It is entirely possible, of course, that the aliens now have to resort to implant technology because their activities are so widely known and too many people are aware of their older methods. If that is the case, it could mean that *they* still need to operate on an individual level – at least for the time being.

While there seems to be no particular demographic bias of implant subjects, a study of which human types are more prone to implants over other groups may be useful. Are all abductees implanted? One might expect more young people to be implanted than older ones, due to the probability of a much longer experimental period being on offer (all things being equal, of course). Fortunately for us, nosebleeds on their own do not indicate implants or alien intervention. However, consistent worrying dreams or irrational fears coupled with other telltale signs of abduction, *together with* nosebleeds for no valid physiological reason, may indicate that the subject has been implanted. Having some knowledge of which human types are implanted may provide clues as to why implantation is performed in the first place.

It has already been pointed out that humans are clearly more physically powerful than the greys. However, if you believe the legend, available evidence does not seem to indicate that *they* are intent on perfecting a 'super grey' – a hybrid super being constructed in a Frankenstein-type way from the 'best bits' of both grey and human. Most of the hybrid 'children' seem to be rather pathetic little creatures who appear to hover between their world and ours. That such physical appearance owes more to a being's environment than its culture is clear.

The greys' strange and almost obsessive preoccupation with our biology may be part of a programme to help the human race adapt to imminent changes in the Earth's environment brought about partly by natural changes and partly by our predisposition to live out our lives in an increasingly polluted world of our own making. However, if the rumours of their physiology is right – one heart/lung organ instead of our two lung/one heart configuration – it might seem to some that the aliens are quite a way behind us in the evolutionary scale. Such a configuration does not suggest adaptation – it suggests a compromise, almost a deliberate design feature. There seems to be little room for redundancy in the aliens' bodies: a serious problem with the one organ would place them at extreme risk, whereas we can survive quite happily on one lung. This might suggest that *they* are rather more of a 'throwaway' item than we are – perhaps a 'cheap' clone copy turned out in thousands?

ENERGY IN CRISIS

It has been estimated that the Earth's crude oil reserves will be a scarce commodity by the year 2050. This is not necessarily because the oil will have run out, but because it will be so expensive to extract from the ground that it will no longer be commercially viable. It is sobering to realize that Western agriculture and food production is so dependent on oil that the lack of it will seriously impact on our ability to feed ourselves. It is therefore obvious that the powerful nations would not rely solely upon the goodwill of the traditional oil-producing countries to provide the most precious commodity for their survival.

Nuclear energy, or rather the nuclear heat-exchange technology to produce steam to run electrical generating turbines, is more of a liability than an asset. While vast quantities of relatively cheap electricity can be produced, this is offset by the enormous overhead costs of providing a safe working environment, together with an escalating problem of the containment of increasing amounts of toxic waste materials, consisting of both low-grade and extremely dangerous and long-lived high-grade radioactive pollutants. The welcome reduction in the convenient production of plutonium for weapons purposes is acknowledged, but the insistence of the Japanese on using plutonium for their fast-breeder reactors and the ever-present possibility of this most poisonous and dangerous material falling into the wrong hands must be cause for serious anxiety on the part of responsible governments – the Japanese included.

Studies carried out in the US by the State Department for Energy have estimated that by the year 2000 some 50,000 tonnes of dangerous radioactive material will be held at nuclear plants across America. The Yucca mountain site in southern Nevada is said to have capacity for around 70,000 tonnes of radioactive waste and will be full by approximately 2030. Also, rather worryingly, from 1980–90 US military waste was accumulating at a rate of around 5,826cu m (200,000cu ft) per year. It would appear that the US is having to carry out controlled pollution of its environment by instituting nuclear waste procedures as part of its Waste Isolation Programme that store highly radioactive sources over 615m (2,000ft) below the surface in New Mexico. It is to be hoped that the geology of the region remains stable.

Far more worrying is the less public domain of Russia and its Republics. The Chernobyl disaster and the environmental damage it caused is well known, but it could easily be duplicated by similar power stations which cannot be upgraded due to lack of either funds or political will. Rotting pieces of ex-Soviet hardware in the form of nuclear reactors from submarines and ships wait for accidents and disasters to happen as they rust away in Northern Scandinavian berths. Regular deep-ocean dumping of highly radioactive waste products (irradiated graphite control rods *et al.*) is carried out by other nuclear powers in Western Europe. The defining problem is – how do you get rid of a dangerous toxic substance that stays lethal for 10,000 years or more? At present we can only bury it in the ground or under the sea: blasting it into space is not an option.

Connect all of these problems we cause for ourselves to natural events such as global warming and the destruction of the ozone layer, and you can see why someone else might be interested in the solution to the many problems if our survival is at stake and life is rare in our Universe.

It should be clear to everyone that the human race is facing a crisis – perhaps not today, but certainly tomorrow. If Bob Lazar's alien reactor can produce vast amounts of electricity, perhaps this is the way to go. The incredible problems of hot fusion, which are akin to the containment of a thermonuclear detonation, and the reputedly arcane art of cold fusion would be obsolete overnight and the US would lead the world in energy production, with all that that implies. Reliance on fossil fuels would be a thing of the past and dangerous nuclear technology could be abandoned forever.

SUPREMACY IN THE AIR

It is rumoured that the US is already using a technology mix in some of its more exotic aeroforms. That does not mean that other, more conventional systems are not being developed in parallel with these. It is possible that the much-rumoured Aurora utilizes both conventional turbines and ramjet technology (the so-called 'scramjet') to propel the vehicle to around ten times the speed of sound at altitudes on the edge of space.

The US spends vast amounts of technodollars in its constant search for air supremacy (but over whom?). For instance, the B1 bomber cost approximately US$278 million per aircraft. Remarkable, especially as the B1 was not considered mission capable! That aircraft was superseded by the B2 bomber, which had stealth capability. Unfortunately, the B2 cost rose to US$500 million per aircraft and the USAF wanted a further US$60,000 million from the taxpayer for a fleet of them. Unfortunately, the B2 suffered from enemy satellite vulnerability, a limited range, slow speed and no defensive weapons systems! Is it a surprise, therefore, that the subsultory alien disc, coupled with an ability to outpace in pure speed anything that existed, attracted the military mind? The downside was that the thing was broadly unduplicatable, useless as a weapons system (not surprising, as it was probably never designed to be one) and, to top it all, the potential US pilots did not fit the seats or understand the navigation techniques!

However, all was not lost. The USAF could probably use the alien technology in their Advanced Technology Fighter (ATF) programme. The alien science would help them develop a system which did not need a human pilot, except to override on-board computer decisions, and would make the biological computer a failsafe or reference point rather than have fallible human beings trying unsuccessfully to make decisions which required thinking speeds far in excess of their capability, no matter how well trained. The next logical step would be to make the on-board biological computer redundant as well. In that event, it would be a battle of machine against machine – economy against economy.

Science fiction? Perhaps – but it should be remembered that the incredible alien technology allegedly available to the US *could* make science fiction into science fact very rapidly. The astounding night-time gyrations of devices at Area 51 indicate that no human pilot could possibly be on board unless normal 'G' forces can somehow be nullified. The devices could be RPVs and no doubt most of them are. However, there seems little to be gained from remotely controlling alien technology in the way one might fly a radio-controlled model airplane, and it is hardly likely that an expansive, elaborate programme would be developed using that method of researching the discs.

MASTER, SLAVE AND SPIRIT

The 'strangeness' of our visitors has compelled many authors to advance theories of their existence and our part in the liaison with *them*. David Barclay's proposal in his erudite *Aliens – The Final Answer* (see Bibliography) that *Homo sapiens* may be a kind of bio-logical construct for the entertainment or use of an extinct intelligent dinosaur species, is interesting. He postulates that human beings may be a product which was devised to perform specific functions – as slaves or as weapons – so that our creators did not have to expend themselves in conflict. This 'dog/master' relationship was presum-ably so well advanced that various types of humans, both men and women, were bred for different duties. If we take the breeding of dogs as a model, it soon becomes evident that theoretically, at least, it might be possible to breed a viable sentient being if the current taboos of human genetic engineering were overcome.

If such a programme had been instituted, a safety mechanism would have been necessary to avoid the possibility of the slaves over-coming the masters. For that reason, an invention called a Limited Life Worker (LLW) may be necessary (my invention, not David Barclay's). The LLWs must have a short lifespan of around thirty years so as to give them no chance to develop a history, tradition or any form of continuity or tradition. New 'slaves' could easily be grown in laboratories, to emerge full-sized and already programmed for the task for which they had been designed. They could not reproduce themselves, so there was no need for reproductive organs or the emotional/physiological desires that go with them. However, there are risks. A mutated species could emerge which might have the ability to live a great deal longer than the proscribed thirty years and eventually, after a great struggle, the masters could be overthrown and the slaves become human – the result being the various races on Earth at the present time.

Science fiction, of course. David Barclay's suggestion that humans are (or were) 'owned' by someone or something is not new and it lies very deep within the human psyche. It is generated out of a constant desire to rationalize our tragic smallness in the awesome face of nature and, with our increasing knowledge, supernature.

The greys' ability to manipulate matter, energy and time frustrates us because it strikes at the heart of our cherished physical beliefs. Scientists refuse to acknowledge that such events are possible, even though they seem to occur with regularity. If the greys can manipulate so much of our world, is it possible that *they* can answer THE question – do we live after we die? According to Bob Lazar, humans have been described by the greys as containers – containers of what? Is it possible that these highly secretive and manipulative visitors have access not only to our physical world but also to the unknown world of the spirit? Maybe that is where *they* originate?

Communication with the spirit world has been attempted over centuries by mediums, soothsayers, tribal witch doctors and many more, sometimes with effect, more often with none whatsoever. Victorian England burgeoned with all kinds of pseudo-scientific mumbo jumbo and many downright charlatans, but most of these efforts peaked out to coincide with the turn of the century. There was great interest in the latest discoveries concerning electricity and it is not generally known that those two great pillars of the scientific community, Thomas Edison and Guglielmo Marconi, worked on projects which aimed to communicate electronically with the dead. As far as it is known, neither of them succeeded. However, there were later experiments by others which did indicate that something strange was going on, even if electronic communication with our dear departed remains unsubstantiated. Remarkably, it has been claimed that discarnate voices have been heard from specially constructed apparatus and, even more remarkably, partial manifestations of form have purportedly been witnessed. But these areas are not even on the fringes of science – they just do not exist at all.

If the greys or other alien entities who visit us do come from other dimensions which are unseen by us, it may be possible to communicate with them across the respective dimensional boundaries. If for one moment you can suspend your cultural beliefs, conventional wisdom suggests that it is perfectly reasonable to consider that Earth is under continual bombardment from subtle and possibly subliminal communication in which we know nothing of the language but nevertheless somehow stumble towards understanding the message. If so, it could follow that the purpose of such unseen 'help' is gently to massage our personal realities, in a transformational kind of way, towards new and better overall human behaviour. This would necessarily take place over a considerable period of our evolutionary time – in fact, for all of our evolutionary time.

Such transmogrifying and ambitious systems might stand the best chance of a successful outcome if contained at a very personal level to avoid a lethal cocktail of massive changes over short timescales. However, there are disadvantages. These may include personal traumas associated with painful and uncertain transitions from one state to another, the uncertainty of trusting in something you do not or cannot understand, and peer problems associated with head-on confrontations with those who choose to ignore the message. There is also the prospect of isolation, which could drive some into severe or chronic mental illness. There would be risks.

Overall, however, this is a more benevolent view of our visitors, and one that a great many people seem to share, probably as a result of the failure of mainstream religion to deliver. Even so, such alien paternalism is very suspect indeed and there is ample evidence to suggest that the real intention of our visitors is loaded with menace – more than most of us realize.

CONCLUSIONS: A NEW BEGINNING?

Truth can only come to people in the form of a lie – only in this form are they able to accept it; only in this form are they able to digest and assimilate it. Truth undefiled would be for them, indigestible food.

George Ivanovitch Gurdjieff

We now need to draw the metaphorical strings tight and review the major known and still unknown factors concerning the aliens and our interaction with *them*. It is true that there are many individuals close to the subject, particularly in the US, who *want* or even *need* to believe it all, the crazy bits as well as the scant corroborated evidence. It seems that the following is a fair appraisal of what is known, mainstream, about the subject.

THE UFO GROUPS

So-called researchers

There are a large number of so-called researchers who still remain keen to log and tabulate occurrences and interview witnesses, and do little else. The broad admission of these ufologists is that most sightings can be explained away by natural occurrences, weather balloons, space garbage, research aircraft or the planet Venus. They steadfastly refuse to admit that intelligent life could exist elsewhere in our Universe because of the vast distances that such life would have to travel in order to visit us. That hypothesis is usually the product of a closed mind, a moderate education or entrenched religious dogma. This group suffers from denial.

Contactees

There are others who consider themselves contactees and believe they have a mission to fulfil on the visitor's behalf: that is, to prepare the way and ensure that the world populations are full of celestial peace and love for the brotherhood of life across the Universe. By and large, this group is harmless. However, these are cult groups who probably still remain useful in some way to the mis/disinformationlists, despite the fact that the 'cat seems to be out of the bag'. This group, by definition, usually wants to start new religions, because they find existing theologies unsatisfactory in the face of their all-knowing 'celestial brothers'.

Serious researchers

Serious researchers will look at and consider all the available evidence objectively. They will not be swayed by emotion or evidence that is not satisfactory. They will have to sift through much pseudo-scientific rubbish and many dead ends to get anywhere near the truth – which of course, remains closely protected, unknown and possibly unobtainable. Self-delusion, or the overpowering 'wish to believe', is a constant danger.

Reluctant witnesses

There are casual or reluctant witnesses who have no real interest in the phenomenon – that is, until they come into contact with it. These witnesses are those who fly airplanes professionally, sail on the seas or carry out other professional duties which bring them into contact with the outdoors. In this category we find police officers, who probably stand a better chance than most of us of a sighting or close alien contact because of their unsocial hours of work. Military personnel also fall into this group.

Government officials

There are various government officials who may know a little of what is going on. Occasionally, ancillary workers will come forward to corroborate a whistle blower's statements. In these situations you either believe or you do not, you don't sit on the fence. Much will depend on whether the source has credibility – would you believe a janitor or a professor? It also depends on whether the story can be checked independently or corroborated by any other person or co-worker who has identical or similar knowledge.

Profiteers
This group sees the subject as one through which they can make money. The profit motive may get in the way of objective research and sometimes even provoke a production-like approach to the subject, as with movie sci-fi blockbusters, wild claims and the odd dubious exposé. While such exposure may raise the temperature beneficially, the waters do tend to become muddied. Subject credibility also suffers.

The public
Then there is the public at large, who would have little or no interest in the subject even if it came up and bit them! This group forms the largest of all and they are usually extremely fearful of any alien contact even if it is just a sighting of an unknown distant object, and more especially if it is a close encounter. Many of this very large group will steadfastly refuse to recognize that the subject is real mainly because it is too incredible to believe anyway. There is a culture problem here.

Abductees
This group may be frighteningly large on a global scale, and no one has yet attempted the daunting task of listing a full analysis of substantiated abductee data for the whole world. It is known that this group is growing in numbers, particularly in North and South America. However, Europe no doubt shares numerical parity with them. It appears that no land mass in the world is ignored, which probably makes Australia and the Pacific Rim countries particularly susceptible due to their remoteness and vast areas of land where civilization is sparse.

Abductees are usually unwilling. Some are repeat victims and seem to know what to expect before it happens. Sometimes routines are carried out to the letter on each subsequent abduction, almost as if the subject is on a regular visit to the out-patient department of a hospital for a course of treatment. However, in the case of a repeat abduction, and unlike in hospital, the treatment is never described to the 'patient', nor is any explanation given for it. In exchange for the use of their bodies, the victims are sometimes given (it is not known whether intentionally or accidentally) a raised spiritual awareness or clairvoyance where none existed before. Others may become psychotic or have relationship problems after the experience and suffer lasting damage. The disturbing incidences of nasal implants (or implants at other sites in the body) seem to be on the increase. While the reason for such implants has already been discussed (see page 208), their *purpose* remains unknown.

There is another important group who have actually been abducted and yet do not know it or have any proof that it happened. However, if the victim suspects that something strange has happened to them and they put the suspicion to the back of their minds, they may cause themselves to suffer from the recognizable symptoms of self-denial or suppressed belief that something like an abduction could happen to an ordinary person like them. This action can leave deep scars in the subconscious, which sometimes surface as recurrent nightmares or bad dreams. In severe cases, psychotic behaviour may develop if the subject is left untreated. These scars in the subconscious are found in equal measure in both sexes due to the deep-rooted memories of their intrusive medical examinations. It is unknown how large this important group might be and it would be revealing to ascertain the demographic concentrations of this category. The fact that the US *appears* to have more of this type of abductee may be misleading inasmuch as the US citizen is generally more motivated to discuss mental health issues than people of more repressed cultures. A very worrying aspect of the abduction scenario is the number of children who are alleged to have been taken. One can only speculate as to what kind of lasting damage may be caused to them by their experience.

The US has the skilled personnel and support infrastructure to assist the individual in their search for the meaning of their dysfunction. The global numbers of abductees who do not yet know they have been taken is therefore probably far greater than anyone realizes. It is perhaps significant that with all the technology at their disposal, the aliens cannot entirely remove the victim's memory of the event. At best, strange screen memories may be incorporated to blot out the incident – memories entirely inappropriate to the victim's life or experience, which usually provide the initial motivation to do something about this mental intrusion. The reason that the aliens cannot entirely remove the memory of their victim may have more to do with our physiology than their lack of technological expertise. Unfortunately for the individuals involved, there are many abductees who claim to have been taken repeatedly in their lives – their experiences are not one-off incidents.

Others

First, there are the dedicated debunkers. In past years, it was generally assumed that such individuals were government agents whose job it was to target and discredit particular individuals and infiltrate UFO groups. While some of this no doubt still goes on, the wheel has turned and the quality of information available to almost anyone who cares to look is far better today than ever before. This tends to make the debunker's job much more difficult. There may also be privateers at work who cannot resist getting involved if they feel that a particular piece of data or information offends their sensibilities. Unfortunately, some scientists fall into this category, although most will avoid the subject rather than risk being misquoted or misunderstood.

There are also those unfortunate individuals who seek fame or notoriety by trying to convince others that they have suffered an abduction when they know they have not. This group need conventional psychiatric treatment. They generally remain on the fringes, trapped in their psychotic fantasies. While the comparatively recent explanation of 'false memory syndrome' or the repressed trauma of child abuse may have some relevance to this group, it is not reasonable to think that such fantasies constructed by a troubled mind would stand up in the face of even half-competent investigation. However, there are growing numbers in this category who must be identified in order to reduce the 'noise' in real cases.

THE ALIENS

It is claimed that the largest group of aliens regularly seen are 1–1.2m (3–4ft), tall, thin, with triangular-shaped heads. *They* are hairless, genderless (at least in terms of conventional human physiology) and have vestigial ears, noses and mouths. Their long, thin arms terminate in four long, bony fingers which appear to terminate in a black 'claw'. *They* do not communicate to us by any kind of speech, but seem to use telepathy. Remarkably, *they* can achieve two-way communication – *they* can read our thoughts as well as transmit their thoughts or replies to us without any effort on our part, even though we are not trained and do not possess any ability in telepathic communication. Their most striking feature is their large, black, featureless oval eyes, which witnesses report as having a powerful mesmerizing influence. They seem peculiarly adapted, or *made*, for specific tasks or duties (cloned?).

They appear to us unclothed and generally in groups. *They* also seem to have the strength (or technology) to support the weight of any size of human being, despite their fragile appearance and light weight (it has been suggested that they may weigh only 23kg/50lb or so). Their skin is the colour of 'dirty pastry' – a greyish shade – and *they* often have an objectionable smell like burnt cardboard about them. *They* seem to have the ability (or the ability to force our minds to *think* that *they* have the ability?) to manipulate time and matter at will. *They* are disdainful or dismissive of human interference and tend either to ignore us while *they* are preoccupied with their business, or may leave an area when disturbed, to return at a later date when things are quieter. They have no regard for the security of even our most sensitive military or commercial areas.

These aliens group are often seen in the presence of taller but similar aliens, but only inside their vehicles by victims undergoing examinations. This larger group, over 1.5m (5ft) tall, have similar overall characteristics to the smaller entities but, according to most witnesses, seem to be in charge of proceedings and appear to direct the human examinations.

The two groups collectively are known as 'greys'. *They* appear to have no need of weapons to defend themselves personally or collectively when in their vehicles.

Other discrete helpers are sometimes seen who are physically different to the greys. The greys and their helpers operate comfortably within our environment, breathing our air and walking in our gravity. This suggests either a technology at work, genetic adaptation or an Earth environment which is similar to their home world.

Alien craft
In the main, the aliens' vehicles appear to be saucer-shaped or oblate spheroids with a domed top, although other vehicles can be strikingly different. All vehicles exhibit incredible light (plasma?) shows,

with occasional coloured or white light strobes. Some devices change colour according to their speed of flight or acceleration. Their method of flight is often peculiar and unlike any conventional aircraft. They exhibit fluttering 'leaf-type' flight with erratic, instantaneous turns and manoeuvres outside the performance envelope of both conventional pilots and aeroplanes. Acceleration and deceleration to and from incredible atmospheric speeds can be achieved while still apparently under intelligent control and with no atmospheric compressibilty effects. The vehicles often respond to light flashes aimed at them from ground sources. To all intents and purposes, the devices are silent except for a low hum and 'whooshing' sounds heard when one is particularly close to a moving craft. Most of these disc-shaped craft are no larger than 18m (60ft) in diameter and typically around 9–11m (30–35ft). There are also much larger craft which appear to remain in high Earth orbit. All craft appear to have the ability to be visible to us and, in some cases, completely invisible to radar. The smaller discs sometimes have the appearance of dull, unpolished or even dirty aluminium – much the same as the appearance of a working jet airliner around the exhaust areas. Landing traces indicate that the discs are extremely heavy and cause deep depressions in soft earth, which is in direct contradiction to their disdain for our physical laws while in flight.

According to Bob Lazar, the disc he saw was powered by an extremely powerful and efficient matter/anti-matter reactor that enabled the Earth's gravitational field to be phase-shifted for flight. While other vehicles may use technologies even more unknown and strange, the flight characteristics of all visitor vehicles seem to indicate at least comparative technologies. Vehicles can apparently materialize or dematerialize, which might be as a result of the working propulsive system or a side effect of some other technology, the purpose of which we do not yet understand. Despite the very advanced nature of these vehicles, accidents do happen and occasionally discs crash, which supports the idea that the craft and their crews are in fact vulnerable.

The discs' propulsive systems appear to be equally at home on land, in the air or in water. Some vehicles can throw what appear to be light beams – blue to capture or immobilize, white to search or investigate – from points on their hulls. Occasionally, witnesses report seeing entities floating up or down inside light beams. Some abductees remember a light beam as being the method of their entry into a disc. It is not known whether or not lights form a communications link from disc to disc. It is likely that the light beams are some kind of particle or plasma energy beam unknown to our science, as beams are occasionally seen to terminate in mid-air. Discs sometimes exhibit an almost playful approach to conventional passenger-carrying aircraft. Their performance seems designed specifically to impress and amaze their captive audience. The aircraft itself is usually unaffected, electronically or physically, by the experience. However, the scaring aspect of high-speed passes coupled with collision-course flight usually leaves a lasting impression on both aircrew and passengers. No clues are available on the navigational technology of the discs.

When interceptor jets are vectored to approach or ordered to chase a disc, they are regularly out-manoeuvred and outpaced. Larger discs have been known to incorporate or 'swallow' intercepting planes, leaving no trace; fortunately, this seems to be a rare occurrence. It is reported that the weapons panels of chasing jet fighter planes are electronically neutralized by the disc, presumably so that they may not fire on it. Sometimes flight instrumentation – engine and navigation controls – is also affected. However, it is not unknown for fighter planes to fire on discs they have been chasing. Whether this is due to a technical oversight or inexperience on the part of the pilot(s) of the disc remains unknown. As far as is known, there is little evidence to suggest that cannon or missile fire has brought down a disc. However, Bob Lazar has mentioned that while he was at S4 he did see one disc with a hole through its rim, as though it had been shot through. Lazar's comment in itself does not constitute evidence that the disc in question was brought down by conventional firepower and he comments that it is remotely possible that experiments were made at S4 to find out whether a disc would be impervious to such treatment. If that was the case, the damaged disc clearly was not! However, as a general rule it seems that the discs *are* impervious to our conventional weaponry when they are operationally active, presumably through the use of the neutralizing effect of a beamed energy field. Nevertheless, it does appear that the discs may occasionally be vulnerable to the Earth's weather systems and some of our early radars (SAGE?). As might be expected, the incidence of their failure due to these influences may now be reduced or even nullified as a result of the aliens' familiarization with these elements and the consequent upgrading of their technology through actual use in the field.

Discs on the ground tend to neutralize the electrical systems of automobiles and other vehicles. Whether this is an effect caused by the propulsive system or simply a precursor to an abduction is not clear. People who have experienced close contacts or a failed abduction attempt have exhibited burn marks similar to intense ultraviolet radiation. In some cases, microwave-type radiation has been responsible for very severe burns. As not all witnesses exhibit these burns, however, it may be reasonable to assume that they occur accidentally or at least unintentionally. Sometimes an abductee's automobile may show signs of the encounter, which may take the form of ash, dust or strange unaccountable markings on the roof of the car. Diesel-powered vehicles do not appear to suffer engine failures on a close encounter as do their petrol-engined counterparts (due to lack of spark ignition systems). However, the lighting systems (if in operation) may fail or flutter with power outages.

It is not unknown for very large discs to upset huge electrical grid systems as well as defence missile batteries. Power outages, surges, brown-outs and other anomalies sometimes occur if discs are seen near such installations. Whether the alien intelligence is testing, demonstrating – making a statement – or merely inspecting is not known.

Sightings

While discs are seen all over the world, specific sightings are reported as being concentrated around areas of military or economic significance such as missile silos, nuclear power plants and hydroelectric generating stations. The discs also seem to be very interested in the geology of our globe, with sightings occurring near or over the deepest undersea trenches and highest mountain ranges. However, that these sightings are few in number may be due to the fact that our seas and mountain ranges are so vast and underpopulated.

Some areas are so concentrated with disc activity that they could be considered hot spots. Such an area is Puerto Rico in South America. This site is particularly important due to the proximity of a US military base, atmospheric research station(?) and the home of the giant Arecibo radio telescope. Other important sites on the American continent include the celebrated Area 51 Groom Lake in Nevada and (possibly) Gulf Breeze near Pensacola in Florida, which is close to a military installation. Mexico, Argentina and other South American countries also qualify as hot-spot areas of the world. In England, there appear to be some hot spots around the Pennines and in North Yorkshire, and also off the North Wales coast. In Australia, the Bass Strait between the mainland and Tasmania has been a hot spot of activity for many years. There are also many more hot spots in countries and seas all over the Earth which are too numerous to record.

While most people now have some awareness of the possibility of sentient life elsewhere than on Earth, they remain unaware of the general seriousness and complexity of the situation. When the discs first appeared in modern times, great efforts were made to debunk sightings as natural phenomena, new military products of either US or Soviet origin, secret Nazi experiments, weather balloons and sometimes even astronomical bodies. The incredible aerial performance of the UFOs soon convinced governments that if the objects were not 'theirs' and were the products of the 'other side', they were in serious trouble indeed. Once it was found that the discs belonged to neither side, nor to anyone else for that matter, these same governments shifted the emphasis and pronounced that the discs presented no threat to national security – that is, they did not *attack* – even though sovereign air space suffered almost daily incursions.

Remarkably, disc incursions into national air space, whether over sensitive military installations or otherwise, merely resulted in a programme of acquiescence by the military authorities who are responsible for our security. The clear message was one of helplessness in the face of these overt intrusions and it should have set alarm bells ringing all over the world – but it did not. Our politicians and military masters appear to be quite content to have discs cavorting at will through our skies, navigating our oceans and parking on our land as long as *they* do not shoot at us. As long as no mass abductions take place, *they* can take individuals when *they* like. Is that what our governments and military leaders would have us think? The backdrop to all of these very serious shortcomings was the even greater secret of a deal and co-operation between the US and the aliens – if you subscribe to the legend, that is.

A massive cover-up, which included the vilification of innocent witnesses, absolute denials that events ever happened and many damage limitation exercises, would soon become part of everyday life. Witnesses would be threatened, cajoled and bribed by those who wanted to keep the whole issue as secret as possible for as long as necessary. Some attempts would be made to embark on a steady programme of information dissemination so that the public would not be traumatized by the sudden knowledge of the alien presence and the shocking admission that their government had been working with *them* for many years. Even more shocking would be the admission that a stage has now been reached where things are running out of control and our alien visitors are in almost total command of the overall global situation. The gradual information release programme never happened. Perhaps it is time that we were all formally introduced to our visitors.

GOVERNMENTS

The Roswell incident may have triggered the first real contact between lifeforms from another part of the Universe and *Homo sapiens.* Denials, cover-ups and subterfuge have clouded this event since 1947. It remains a matter of personal judgement whether you believe that the initial cover-up was instituted to save the face of the US military establishment over Project Mogul (see page 23) – subsequently blown up out of all proportion by the UFO community – or that a live alien was captured, or that autopsies were made on the dead aliens found at the crash site, or that the damaged disc, aliens and other scattered wreckage was taken back to Wright Patterson Air Force Base. But all this is immaterial. What *is* important is the fact that *something did happen at Roswell.* All the circumstantial evidence points to a cover-up of gigantic proportions and one which cannot justify the attempted character assassination of USAF personnel.

Such a massive cover-up would surely have been unnecessary if the subject of the crash was merely the wreckage of a radiosonde balloon, no matter how advanced in its construction. The other 'official' rumour, suggesting the wreckage was that of a test-fired German V2 rocket, does not have much credibility due to the description of the material found on the site and the testimony that humanoids were involved. We cannot delve too deeply into the subject of Roswell here, and there are many specialist books on the incident to consult for further details. However, it is important to grasp that Roswell may have been a turning point for the US and its relations with the aliens known as greys – always assuming that the balloon story was a cover.

The US government does not have a monopoly on the issue of UFOs. In 1974 the French Defence Minister, Robert Galley, took part in radio interviews which included witness reports and statements from three eminent French scientists, one of whom was astrophysicist

Jacques Vallée. Monsieur Galley tacitly admitted that the subject should be approached with an open mind and refused to dismiss the observational data presented out of hand. Prior to the break-up of the Soviet Union, Russian scientists had been even more effusive and certain of their ground in confirming the belief in the ET hypothesis. The UK government, however, has been deafening in its silence on the subject, despite the British 'Roswell' which occurred at the Suffolk US Woodbridge Base in December 1980 (see page 20).

Is it credible that the greys would wish to do business with several different nations at the same time? Surely, it would be better for *them* if the communications conduit were confined to the most powerful nation on Earth, which at the same time rather fortuitously possessed the right kind of non-converging, dispersed social infrastructure in which to undertake covert experimental programmes over a long period. This kind of behaviour would require very close co-operation from both parties and a contract would be drawn up to define the rules.

The deal

We can only imagine what such a treaty would have contained. Let's make an attempt by considering what we might want if *we* were the ETs. Firstly, we should want security; that is, we would want our hosts to protect our small group from any intervention or disturbance by the remainder of their kind. After all, we would only want to deal with one group of 'aliens'. This implies a very secure and secret headquarters which would be unknown to all but a very few of our hosts. We would want equipment, facilities and power. We would certainly want food we could eat, water we could drink and so on. We would want secure communications so that nobody could intercept our radio transmissions. We would want access to Earth scientists or specialists to help us understand our hosts and their environment. We would want our hosts' agreement that we could take some of their flora and fauna for experimental purposes. It would be essential that our hosts were as technically competent as possible, otherwise we could not communicate with them on any meaningful level. Therefore, our hosts must have already achieved a minimum standard of scientific and intellectual competency before we can even consider any form of useful contact.

In return, we might give our hosts some of our technology and science. However, we would be very careful not to give them anything which could be adapted as weapons in case they might be used against us. Another reason for not handing over our most powerful science would be to avoid giving a technically emerging species the power to destroy itself even faster and more easily than it could with its own thermo-nuclear weapons – that would, after all, negate our purpose for being on the planet.

Our hosts would be elated and wild with enthusiasm about our involvement with them and the new opportunities that our science would bring, but they would soon realize that they could never tell their public about the liaison for fear that such knowledge would destroy their credibility: they had lied about our presence for too long. There would also be a terrible risk of the world's nations splintering into factions, with each one targeting our hosts as the enemy of the world. That could start a world war, which we know is possible given the enormity of the situation. We would understand our hosts' plight almost from the beginning of our liaison with them and would retain this leverage for use at a later date when we might need it. We would be very careful not to interfere in our hosts' affairs, nor would we intervene to save them from disasters, natural or otherwise.

This arrangement might have remained effective for many years. However, other nations eventually learned about us and our hosts found it necessary to renegotiate a new treaty which enabled the expanded circle of contacts access to us. We were not particularly happy about this because there was nothing in it for us. After all, we could take what we wanted, whenever we wanted, so we had no need of further bases anywhere else. The new group sought to gain technical advantages over our original hosts. We thought this counter-productive to our mission and we did not wish to get involved in another species' political rows. We told our original hosts that we did not want this arrangement any longer and if co-operation was to continue they should get rid of the new group, otherwise we would break off the liaison and cancel the treaty.

Our original hosts concurred, but not without some rancour and bad feeling. We did not know or care about any residual agreement that existed between them and the other groups of humanity, but we were now very suspicious of our hosts. We were amused that they seem to be so complicated socially, given that our race system is so simplistic when compared to theirs. After the débâcle of the proposed new group, we decided to treat our hosts less gently. We carried on with our random and systematic examination of their kind (in accordance with their original permission), together with our research programme of genetic adaptation and hybridization. By this time we had lost interest in their technology, such as it was. However, we did warn our hosts that the crude technologies they were using could, and would, damage their ecosphere. They knew that we had augmented our programmes and asked us to be more careful in our operations, otherwise they would not be able to keep the lid closed on our presence. We considered that we did not need to comply. Instead, we suggested that our hosts let their people and the other peoples of their world know that we were on their planet, as part of a gradual information release programme. In that way, their world population would not be traumatized. They agreed to the principle, but could not see how it could be done without the total collapse of their own power system and of societal structures around the world.

At this stage we lost patience, and forged ahead with our plan for hybridization. If we could perfect this technique so that 'our' humans were indistinguishable from theirs, we would have no further need of our hosts, as we planned to place our hybrids in powerful positions within their societies which would be of benefit to us. It would not be so much a takeover as a merging operation. There were not that many of us in the first place and we now knew the full extent of the human population; while it was clear that we should be at a distinct disadvantage living openly in the human world, this would not be the case when we had achieved our merge programme. We could then start to redefine the Earth for our purposes as a sister planet to our own beautiful but different home world.

So, could it be that *they* are constructing hybrid races in a galactic colonization programme – a new but non-human evolution to spread sentient life across the myriad systems in our Universe? If that is the case, we are likely to stay Earthbound and contribute to the Grand Plan only by being 'farmed' for our genetic material. If that is their goal, there is clear significance in the fact that a hybrid human/alien entity has been chosen. It is, after all, quite logical to seed different but essentially related species across the Universe, as no one race will look upon itself as elite or unique. The fact that races will be related allows a structural brotherhood to form which could reduce the risk of damaging conflicts, especially if everyone knows that life exists on a particular a planet. In *Secret Life – First-hand Accounts of UFO Abductions* (see Bibliography), Associate Professor David M. Jacobs states that a very conservative estimate of US abductions reveals that possibly over a million people have been abducted – one wonders at the total number worldwide. Professor Jacobs has also zeroed-in on why the grey nation is here: to raid our human genetic material. He maintains that, based upon the available evidence, there can be no other reason.

Is all the above pure fiction? These are just two possible scenarios from many specifically concerned with the greys and their presence on Earth. There could be many other scenarios, each as fantastic and 'out there'. The above 'probables' are based on rumours that have been around for a while, and they are still being developed. The claim that certain intelligence community elements within the US considered putting on a travelling show called 'Cosmic Journey', with UFO photographs and pictures of alien autopsies (details are given in *Alien Contact – Top Secret UFO Files Revealed* by Timothy Good – see Bibliography), suggests that the US government may have had plans for a trickle release of information. Inexplicably, the Cosmic Journey project seems to have been shelved in early 1990 'for financial reasons'. Is it likely that such a serious and crucial undertaking would be halted simply for lack of funds? The truth behind the cancellation might more profitably be sought elsewhere. At the risk of displaying considerable paranoia, could it be that such a weak attempt at disclosure as Cosmic Journey is aimed more at securing public sympathy for an emasculated administration (which would, of course, be too little, too late)? The *real* truth must soon be told by our respective governments, even though we may not like what we hear. It is up to each of us to seek out that truth and cause our administrators, keepers and protectors to tell us what is known of the alien presence on Earth. Only then will we be equipped to understand better the nature of the wonderful Universe in which we exist or, conversely, the serious threat that is posed to us by the aliens' presence. The acquisition of this knowledge would be the greatest event the world has ever seen – greater than all the religions, and greater than all our achievements and inventions. But we should still be told.

THE HUMAN CONDITION

Without the knowledge of the existence of other sentient life in our Universe, humankind appears to live alone in programmed construct. Not only is our world predetermined, but the Universe itself gives the impression that it too is hardwired. Human beings have the temporal disadvantage of an intellect, which causes us to investigate, deduce and wonder. Our intelligence isolates us acutely from the rest of mute nature. While our reasoning powers are great, they are at the same time tragically wasted in the space of our own short time. Descartes once said, 'Man is an animal with reasoning power'. If he was right, then we may not have a spiritual dimension. The spirit may be a chimera, a self-deception or perhaps part of our evolved programmed biological response to a Universe as cruelly indifferent to our emotional needs or demands as are our visitors.

The programming of our world is attributed to nature or, more correctly, God. It involves a reliance on interwoven dependencies: male/ female and plant/animal. There are varieties of tree which depend completely on a particular species of bird for their survival – or is it the other way around? Did the bird invent the tree? Did the pebble invent the mountain? All life must be equal in importance, for it shares a universal consciousness. There is no *real* distinction between one form of life and another, they are all interlinked and dependent on each other. Matter belongs to matter, sentient or otherwise. The mightiest of mountains is dependent on the tiniest stream and the featureless wind. The smallest insect is dependent on the tiniest flower, which in turn is dependent on the insect, so that each short life may occur. We are, unconsciously, part of an enormously interwoven and complex system of interdependencies.

Human beings straddle a very thin line between the incomprehensibly large and the incomprehensibly small. Our given senses limit us deliberately, by restricting us to close parameters in which we see, hear and touch the merest whisper of creation – we cannot even hear her voice. Our instruments cannot detect all of creation – even if we knew what creation really is. Human beings can be incredibly violent and at the same time wonderfully saintly. We are surely a developing species which has been caught in the act of emergence from its chrysalis of perception – even now, still only vaguely aware of other realities and possibilities through the miracle of our imagination and resolute, stubborn will to define and understand.

The visitors' presence on Earth is far too important to be kept secret *for any reason*, whether it be for military purposes, from an outdated nationalistic mind set or as an excuse to avoid the embarrassment of a government which has to admit publicly to the world that they have been host to *them* for fifty years. That the evidence suggests the aliens are content to interact with humankind on an individual rather than a global level indicates a timetable of events which must stretch well into the next millenium. It also indicates the terrible impotence of our governments to do anything about it, presumably because of the crushing effect of a disclosure. If the aliens' presence on Earth was beneficent, we should have known about *them* almost from the beginning. The fact that we have not should be considerable cause for concern for everyone.

It would be wise to remember that we have lived with angels and demons in the past; our religions and histories are littered with them. However, our current demons seem much more tangible and insidious than before. Perhaps now is the time in our history when we collectively need to feel the demon's breath before we may experience the tender touch of angels. We should rightly be fearful that we are individual prey, for that means we are all essentially alone and no one can protect us from the visitors. If any government has opened the door to these unwelcome guests without our permission, we have the right to know about it – NOW.

APOCALYPSE?

A final thought: what if the apocalyptic vision of an exploding Earth shown to an increasing number of abductees is not simply a show put on for effect? Could it actually be a true vision of our inescapable future? Can we just shrug off Mayan predictions of the end of the world on the 23 December 2012, or Edgar Cayce's warnings of a shift in the poles of the Earth resulting in massive upheaval and destruction on or around the turn of the century? Should we brush aside the fact that on the 5 May 2000 there will be a conjunction of five planets – Mercury, Venus, Earth, Mars, Neptune and Uranus – which could result in some gravitational effects being felt by us? Should we ignore the prospect of being hit or near-missed by a comet predicted to arrive in the vicinity of Earth at around 2026?

If *they* are attempting to warn us of an impending global natural disaster, we seem to be ignoring the message because we do not understand their logic. Alternatively, it may well be that their accelerated schedule has more to do with getting their job done before we are destroyed. On the available evidence, it is difficult to believe that *they* would wish, out of some strange galactic altruism, to intervene to prevent our destruction (even if *they* could). If both occult and scientific predictions of natural disasters are wrong and we are destined mutually to destroy ourselves in nuclear fires, it is reasonable to think that *they* would not wish to get involved in the conflicts of a totally different species, with obvious risk to themselves, for no return – after all, would we?

Should we take the oft-reported 'love' that flows from the tall insect-like grey alien in an encounter as being the same love that we know in our human relationships? Or is it a contrived love with the purpose of getting what they want from us with the minimum of fuss? When we love, really love, we look beyond what we see and choose not to notice the imperfections in our partners. We tend to overlook a person's bad points. Instead, we seek to know the persona, the vivacity and the soul, the real person behind the familiar mask of the body. In that way, we can form lasting bonds between one another that some-times only death can break. Our visitors do not exhibit these familiar human qualities and frighten us with their appearance and their dirty, smelly, examination rooms.

By definition, *they* cannot share in our lives or our human purpose, weak as it is when compared to their superior technology. *They* appear not to understand our emotional and spiritual needs and treat us as we might treat an animal under medical examination: with firm kindness and implacability. *They* do not care about our essence, our individualism, and that is their weakness for, paradoxically, it is our frailty which makes us stronger than them. We have such a short time in our lives to complete the great works that persist despite us and our frailty – we are so often led astray by our emotions and get things ter-ribly wrong. It may be that the only thing to keep us safe from *them* will be our indomitable human spirit. If we feel their false love, we should ask ourselves a simple question: 'Is what I feel matched by their actions in this encounter?' If the answer is 'No', it is not love we are being given but a subterfuge to achieve an end, and it is there just to make us more pliable in their probing experiments.

LIES, TRUTH AND KNOWING

Unfortunately, there are apparently some people who would try to outdo even the most malevolent alien. The incredible and fright-ening claims of John Lear and Bill Moore of aliens keeping human body parts in amber liquid and of dreadful experiments being carried out on captured humans seem to be aimed squarely at those in the UFO community who want to believe everything that is presented to them. This also discredits those serious researchers who have their feet on rather firmer ground and is extremely dangerous for the mental health of those who have left all reason behind. Such statements may even be designed to fracture the UFO research community, thereby causing lasting and irrevocable damage. The recent fiasco of the Santilli Roswell autopsy footage is becoming typical of what is happening, with claims and counter-claims circulating at the same time within the UFO community.

The cottage industry that has centred around Gulf Breeze near Pensacola Bay in Florida because of the publicity surrounding Ed Walters (see page 17) is also cause for concern. The fact that it is alleged that Bruce Maccabee, an optical physicist, had been *paid* to examine Ed Walters' Polaroid photographs tends to call into question the objective nature of that examination. There is much speculation over Polaroid photograph 19, which shows an unidentified vehicle hovering over a road at night, and which a frightened Walters purportedly took from the safety of his pick-up truck. Bruce Maccabee claims that the object is about 4m (12ft) in width and about 3m (9ft) in height. He also deduces that it is about 1m (3ft) off the ground. He mentions that the width of the so-called 'power ring' is around 2.3m (7½ft). The 'power ring' on this particular picture has been the subject of much discussion since it was published and several photographic analysts have questioned the authenticity of the shot because the 'power ring' does not cast the expected signature on the road – something looks wrong. The close proximity of a substantial naval installation seems to have had no impact on the 'believers', and the fact that the Ed Walters case eventually dissolved into the familiar hoax routine has probably not deterred the nightly skywatchers who eagerly await the lightshows.

The more recent *Guardian* video footage of a supposed UFO landing at night near Ontario, Canada, which was sent to mainstream researchers in the early 1990s also caused a minor stir within the UFO community. Having seen the footage myself, I can say that the only thing that was slightly impressive was the sound of a dog barking in the distance and the breathing of the cameraman in an otherwise totally soundless environment – even when the object seemed to take off. The footage of an apparently luminous familiar 'alien' face with luminous hands, remaining motionless and posed for the camera, adds a somewhat surreal effect. The fact that modern digital techniques can create dinosaurs which roam over the countryside in broad daylight eating everyone in sight, or battles fought in deep space between opposing intergalactic races, seems to have been ignored by the believers, who are unable to understand that with the right equipment it is comparatively simple to dub a soundtrack on to a silent video. Readers may be interested to know that allegedly MUFON maintained that optical physicist Bruce Maccabee hoped to make money from the story, something which Maccabee has subsequently strenuously denied.

What do we make of all these sensational claims? Well, it could be that some smart people have found out how to make a lot of money selling such stories to an eager UFO community, hungry for information and sensationalism and convinced that there is some kind of massive global cover-up, orchestrated by powerful governments in the pursuit of their own nefarious agendas. On the other hand, it could well be that the claims are largely true: if so, it is clear that we are all in very serious trouble. Could these stories represent an attempt at disinformation and misinformation spread by the perpetrators of the 'cover-up'? That would mean legions of government agents operating all over the place. How do you keep control of that kind of situation?

Nevertheless, we *continue* to see strange objects in our skies, we *continue* to find quite ordinary human beings claiming that they have been abducted by creatures from another place and we *continue* to ask questions of governments who stonewall by telling us that these objects have no defence significance! Something is going on, of that there is no doubt. Even the most sceptical among us must surely ask the question: if these strange objects we constantly see in our skies, on our land and in our seas are not ours, whose are they? If we can ignore for the moment the scientific gaps and inconsistencies in Bob Lazar's statements about working on alien disc propulsion technology for the US government, and the reasonable but somewhat irrational hypothesis that he may have invented his story to 'get back' at his employer or perhaps the 'system' itself, we still do not come up empty handed. There is just too much first-hand and circumstantial evidence to dismiss it all summarily as nonsense. Someone, somewhere, knows what is *actually* going on.

The continual 'noise' generated by the so-called 'UFO community' does not help in the search for the truth. Nor do the occasionally crude attempts at debunking carried out by those hostile to any idea of the ET hypothesis, from either conviction or just plain bigotry. That most sightings of strange things on and around our planet are comprised of re-entry junk, the odd secret and not-so-secret project, mistaken identification or just plain old-fashioned hoaxes is not in question. Such things will always be with us. Perhaps our failure to understand the phenomenon scientifically has more to do with a fashionable mind set, which assumes it is reasonable to reject an anomalous one per cent of the result of any experiment as being irrelevant in the face of a 99 per cent predictability. Perhaps it is the one per cent we should be concerned with. The inescapable truth is that there is *no* scientific paradigm for a real UFO event.

In the unlikely event that there really is no cover-up and our respective governments are telling us the objective truth, it must be reasonable to assume that there *should* be a cover-up of massive proportions while people all over the world continually record sightings and contact with the visitors. If these objects do not originate from our own technological efforts, they must by definition originate from somewhere else – and from *someone else's* technology. Cover-ups may be generated out of the arrogance that is sometimes displayed by governments who appear impotent in the face of the phenomenon, and who cannot admit their impotence publicly for fear of political instability. The much-used phrase 'The phenomenon does not constitute a security threat' is frankly no longer acceptable: a transient unknown object that intrudes into *any* sovereign air space without permission *must* have security implications. Governments can hardly claim a lack of data, as it is known that their agencies and departments have been collecting information for decades. In the unlikely event that there is no official cover-up and our leaders are as puzzled as we are about the visitors, we should all be extremely concerned about the ability of our respective armed forces to protect us. The reality which is now gradually being uncovered suggests that our leaders are not as ignorant about the visitors as they would have us believe.

Judging from all the information available, it would be unwise to ignore the visitors' presence. It is crucial that any visitor event should be tested by remaining objective and sceptical even in the face of what on the surface might appear to be conclusive evidence. It is extremely unlikely that conclusive evidence will exist in the public domain, although many believe it does exist in secret government archives. Conclusive evidence would be priceless – if it exists in a form we can understand. Unfortunately, there is a natural tendency among the various UFO groups to want to own UFO data, photographs, documentation and so on – anything which can be considered to be circumstantial 'evidence'. This tends to hamper serious research by making the data exclusive to one group alone, which does not enhance the prime objective – to learn more about the UFO phenomenon by the free sharing of data and the cross-fertilization of ideas which unbiased opinion can generate.

There are pressing and urgent reasons for disclosure by all world governments of the truth about these unwelcome visitors to our world. It is up to each and every one of us to seek that truth by continually asking searching questions of our leaders. We should all be cautiously afraid of the visitors, whether *they* come from the compact 'Multiverse' proposed by Jacques Vallée or from star systems many light years from Earth. We should be equally concerned about the stimulation *they* bring to us and perhaps the almost imperceptible evolutionary jump caused – in the absence of the hard evidence which is currently denied us – just by their *suspected* presence.

Finally, if we survive the prophecies of doom and somehow come to terms with the fact that all humankind has to live on Earth in harmony, perhaps the philosopher or the liar poet will have the last word. What marvels will be left when science knows everything that can be known? What mysteries will there be to solve when we know the entirety of God's plan? For then the lies and truths of humankind will merge into a knowing which has no bounds. By that time we shall already have outgrown our beautiful Earth and will have joined the star people in the Great Universe – our true home.

EPILOGUE

As I write these closing words, the NASA Martian lander is exploring the surface of Mars. This technically stupendous feat proves that NASA can still 'cut it' when they have to. I know the question you are asking: 'If ET is already here, why is the US appropriating billions of tax dollars to search for life in the near planets?'. The answer to that question may involve one of the following:

PROPOSITION 1

NASA scientists are kept totally in the dark about ET already being here.

Probability

This is very unlikely. Most NASA scientists are aware to a greater or lesser degree of the rumours and legends surrounding current ET claims. However, as their projects (and their salaries) are bankrolled by the US government, there must be a political element which will determine the agency's overall policy and the public face it portrays in such matters. NASA scientists will therefore be subject to the predictable 'safe' scientific response: 'ET does not exist because we have not found any evidence yet.'

PROPOSITION 2

A question: why didn't NASA send their Martian lander to the Cydonia region of Mars, where it seems that a giant 'face' and peculiar pyramidal structures may be discovered?

Probability

NASA are on record as stating that in their opinion, based upon previous orbital photographic surveys, the supposed structures are tricks of the light and there is no evidence to suggest that anything would be gained by sending a mission to that area of Mars at the present time. However, there is a possibility that such a trip will at some time in the future be privately funded by a conglomerate interested in discovering what really is down there on the surface. Has NASA deliberately been told to keep away from that area of Mars? If it has, was it by the US government? Or someone else?

PROPOSITION 3

NASA is an incredibly expensive red herring, designed to keep attention away from the real issues.

Probability
This does not seem a viable proposition. However, there are a great many spin-offs for US industry and technology from the general space programme which are not directly connected with NASA 'spectaculars'. It should be remembered that NASA tends to float on the backs of many small companies, which in some cases amount to not much more than garden-shed workshops run by individuals with extremely specialist knowledge and a high degree of skill. If NASA was set up as a diversion to the truth, it is a terrifyingly risky business to undertake, given the ever-present possibility of a mistake or someone blowing the whistle.

PROPOSITION 4

NASA is an extremely valuable scientific and political institution which serves the US both at home and abroad. Scientists are detached from projects other than their own specialist areas, even though their expertise could be useful elsewhere (S4?). A Chinese Wall therefore exists to separate NASA scientists from very sensitive areas which are of no direct relevance to their immediate projects and goals. Methods which use ridicule or NASA project experience to discredit the ET legend are actively encouraged by the administration.

Probability

Of all the propositions listed, number 4 seems the one most likely to be correct. The ability of the US government to use NASA as a means of fostering or regenerating national pride is clearly very valuable. As long as NASA maintains a successful outcome to its endeavours, it will no doubt prosper. However, NASA is not just a powerful domestic tool, it is also a direct scientific line to the US economy, as other countries are discovering. International space exploration in the spirit of friendly co-operation between the nations that can afford it will be essential to the US economy in the future. The nations who cannot afford to join this exclusive club will, quite literally, be left behind.

This predictably meritorious, capitalist approach cannot take into account the darker side of what is really going on behind the scenes, even as we are being globally beamed the daring exploits of the brave astronauts playing out their finely rehearsed charade for us on our televisions.

However, we should not seek to diminish what NASA or any other space agency is achieving or has achieved in the past, for they have done so through much adversity, effort and, sometimes, sacrifice. Nevertheless, it is reasonable to ask why we have not achieved so much more with co-operation from our visitors, for there appears to be no evidence to suggest that any outside help has been given to *Homo sapiens* in our tentative exploits in space. The fact that we have not received any help from *them* does not, of course, mean that *they* do not exist. On the contrary, in the face of so much evidence to support the contention that ET is already here, it suggests that the time for such co-operation may be long past – if an opportunity for co-operation ever existed in the first place.

The recent strange admission by the US government that the USAF lied about UFO sightings in the 1950s and 1960s in order to protect very sensitive spying operations using the Lockheed U2, flying at altitudes of up to 21,945m (72,000ft) and with a range of around 5,920km (3,700 miles) in its later marks, does not really add up. The missions undertaken by the U2 were the usual overflights of Soviet territory, with the emphasis on mapping missile sites. The plane was also used to overfly Pakistan and Norway on spying missions. The U2 appeared operationally in 1955 but was probably used prior to that date. Almost parallel to the operational use of the U2, the USAF used modified British Canberra aircraft with high-aspect ratio wings to overfly Vietnam in the mid to late 1960s in recce missions.

The other aeroplane mentioned in the USAF admission was the Lockheed SR-71A, code name 'Blackbird'. This plane generically replaced the U2 and its mission capabilities included a operational height of over 25,000m (82,000ft) – the altitude record stands at 26,175.074m (85,068.997ft), Capt Robert C. Helt, 28 July 1976 – with speeds up to Mach 3 and a range of almost 5,500km (3,500 miles). About thirty SR-71As were built and nine planes remained operational in 1983. The fact that the SR-71A remained operational in 1983 (and beyond?) implies that the USAF are still under instructions to lie about sightings that occurred way into the 1980s. There remains much speculation about the SR-71A successor (Aurora?).

Quite how the US government hopes to explain away UFO sightings on its own territory, featuring strange circular vehicles behaving in a distinctly un-aerodynamic fashion and capable of accelerating from stationary mode to thousands of kilometres per hour in the blinking of an eye, and have these explanations believed, is frankly breathtaking. While some people who saw the contrails of these high-flying spy planes may have thought that they were witnessing something otherworldly, the vast majority of observers would have noticed the difference between a regular aircraft and the flying patterns of a UFO. In any event, as the spy planes operated over specific areas of the world, their presence would be seen by the same people time after time. (As a passing comment, the year of 1997 marks three particular fiftieth anniversaries strangely juxtaposed to one another: one, Kenneth Arnold's sighting of strange boomerang aeroforms 'skipping' like pebbles over a pond in the vicinity of the Cascade mountains of Washington, which seems to have started the modern wave of UFO events; two, the celebrated Roswell affair; and three, the creation of the USAF.)

It is not the admission of the US government that lies were told to protect sensitive national secrets which is particularly interesting, rather it is the timing and the *need* to make such an admission in the first place. Could it be that researchers are getting too close to the truth for comfort? Or is it possible that as we approach the millenium the US government is getting ready to tell the world what we have waited so long to hear?

While we wait for the announcement, the cattle mutilations in the southern states of America continue unabated. The almost silent black helicopters continue to range over these areas with their cargoes of oriental-looking crew. Angry ranchers who have lost cattle will no doubt continue to shoot at the silent craft, frustrated because they cannot identify those who surgically core parts and organs out of their valuable livestock, leaving them mutilated and lying almost bloodless on the range or in corrals, and because their government will do nothing except tell them that it is the act of predators and offer the farmers compensation for the loss of their animals, when the farmers know that no known predator could inflict that type of horrific injury on their beasts.

While we wait for the announcement, the *chupacabra* (goat sucker) continues to rampage over an increasingly wide area. First reported in Puerto Rico, then large areas of Central America including Mexico, incidents are now said to be occurring on mainland Europe, in Spain and Portugal. It seems the creature is breeding and occupying territory at an alarming rate.

This strangely elusive and visually frightening creature appears to be totally unknown to our biology. While wary of people, it dispatches its prey using what appears to be a long tube-like tongue that bores deep into a sheep, goat or other similar-sized mammal to suck out the animal's blood, leaving it almost empty of body fluid. In the process, the *chupacabra* destroys internal organs. This nightmare of a beast seems more at home in a cheap novel of the science fiction genre than in our reality. Perhaps it is an escapee from a genetic laboratory, or a beast which was brought by our visitors from their world – to be unleashed on ours.

When all the gaps in our *true* history are filled, our heritage may not be the one that we cherish. Instead, it could be one that would make us more like *them* instead of the way we are, supposedly a product of simple biological evolution.

It is possible that we are *compelled* to be the way we are because we just happen to live on planet Earth and for no other reason, in much the same way that the denizens of another planet light years away from Earth who may exist merely as intelligent slime have the same right as we do to call themselves kings in their world.

This scale of adaptation by universal nature to produce life from the available materials is wondrous indeed and we should be humbled in the face of it. Ultimately, such acceptance that there *is* something bigger than ourselves out there may help us to overcome our fears and prejudices, and confront the dangers ahead with confidence.

Raymond A. Robinson
West Sussex

If you want to contact me about the contents of this book, or if you suspect that you may have been abducted or implanted by aliens, please write to the publishers, who will ensure your letter reaches me. I will endeavour to reply to everyone, but it may take a little time, so please be patient.

APPENDIX

One light year = 9.46 million million kilometres
Light speed = 299,792.5 km per second (in vacuum, i.e. space)

RETICULUM

The Reticulum is a constellation introduced by Lacaille in the 1750s. The name is supposed to commemorate the 'Reticule' astronomical instrument which he used for measuring star positions in the southern sky. Reticulum is not a notable constellation and lies near the large Magellinic Cloud. The three stars are:

Alpha Reticuli – a yellow giant 390 light years away.
Beta Reticuli – an orange star 55 light years away.
Zeta Reticuli – a double yellow star, similar to our Sun, 40 light years away.

CONSTELLATION LYRA

The constellation Lyra contains the fifth-brightest star in the sky, Vega. Our Sun is carrying the solar system in the direction of Vega at a velocity of 20km (12½ miles) per second (relative to near-by stars). Because of the phenomenon known as 'precession', Vega will become the Pole Star in about 12,000 years' time. Meteor showers which visit Earth emanate from Lyra, one in April and one in June. The visible April shower numbers about twelve per hour while the visible June shower numbers about eight per hour.

CONSTELLATION PLEIADES

The Pleiades form the brightest star cluster in the sky. A total of about 250 stars make up the cluster, which is about 415 light years away. The Pleiades formed in the last 50 million years or so and are therefore relatively very young. Many blue giant stars with surface temperatures in the region of 20,000°C are in the cluster and a great deal of nebulous gas still surrounds the system (our Sun, a typical main sequence star, has a surface temperature of around 5,500°C). (See photograph 8.)

THE GALAXY

In common with many other galaxies in the known Universe, our galaxy is spiral in shape. Arms radiate from a central bulge or body of stars. It is estimated that the galaxy is about 100,000 light years in diameter and our Sun and its system lie in one of the arms, about 30,000 light years from the centre of the galaxy. Scientists have estimated that the galaxy contains about a billion stars.

THE SOLAR SYSTEM

In our solar system, all the planets tend to orbit the Sun in the same plane. The planets rotate elliptically around the Sun in an anticlockwise direction. Venus has the most circular orbit but, interestingly, rotates on its axis from east to west, the opposite way to Earth and the other planets. Unfortunately, it is now known from probes that the surface and atmosphere of Venus could never support life as we know it.

THE ASTEROID BELT

The notion that the asteroid belt represents the remains of a destroyed planet is not substantiated by the fact that even if all the 2,000 known pieces of rock and debris were reconstituted, they would still make up a body under half the size of the Moon.

BOB LAZAR

Bob Lazar mentions that he saw a black hole appear when the reactor is working and the gravity amplifiers have been focused to produce an intense gravitational field. Conventional physics can define at what point massive planetary or stellar objects become a black hole. The formula used is called the 'Schwarzchild radius':

$$R = \frac{2GM}{C}$$

where R = Schwarzchild radius, M = mass, G = gravitational constant, C = speed of light.

Using the formula, the Earth would become a black hole if compressed to less than 1mm (⅖in) across. The Sun would become a black hole if compressed to less than 1km (⅝ mile) across.

HOLLOW EARTH

Just prior to the rise of Nazi Germany, a group calling itself the Berlin Lodge (or the Vril Society, interwoven with other Theosophical and Rosicrucian groups) believed that a 'super race' of people lived deep underground near the centre of the Earth. When the time was right, they would emerge to change the German race, making them as godlike as themselves and thereby enabling them to rule the world.

THE MONTAUK PROJECT

This is a bizarre story in which it is maintained that the US government has carried out covert experiments at the old USAF base near Montauk Point, Long Island. The base was officially closed in 1969 but allegedly continued to operate without any apparent US government sanction or funding. Local legends persistently claim that the 1943 experiments to attain radar invisibility of *USS Eldridge* (the so-called Philadelphia Experiment, in which Dr John von Neumann and Nikola Tesla, among others, are said to have participated) culminted in the Montauk Project. The project was said to have stumbled on a microwave mind-control system and a method of time travel through the electromagnetic manipulation of space–time. The project was said to have been shut down in the summer of 1983, when it had allegedly become extremely dangerous and was running out of control. A sinister aspect of the story concerns the supposed physical transportation of countless numbers of people through a time portal to a date in our distant future. (Details are given in *The Montauk Project – Experiments in Time* by Preston B. Nichols and Peter Moon – see Bibliography).

THE TRANSISTOR

A 1967 reprinted version of the *Penguin Dictionary of Electronics* by S. Handel describes a transistor as 'An active semiconductor device with three or more electrodes'. The dictionary goes on to list the following transistor types: bipolar, conductivity modulation, filamentary, hook, junction, N-P-N, P-N-I-P, P-N-P, point contact, power, surface barrier, unipolar transistor, amplifier, complementary amplifier, pentode and tetrode.

It is not understood why the surface barrier transistor – designed for operation at frequencies up to and above 100 mc/s, produced by accurate electrochemical etching and plating techniques which allow extremely thin barriers to be used in the semiconductor – was singled out from all the others as a product benefiting from alien intervention. However, its very high operating frequency and method of fabrication may have isolated the SBT as a unique device of its time and of particular use to the visitors.

A SEA OF CONSCIOUSNESS

Pierre Teilhard De Chardin was a Jesuit philosopher-scientist who speculated that the Earth had an evolutionary envelope, which he called the 'noosphere', that determined the future – just one of several extant spiritual explanations for humankind's collective evolutionary progress in the world. Theories range from Earth's entrapment of human beings' evolutionary psychic energy, i.e. consciousness (the law of the conservation of energy states that energy cannot be destroyed, so if psychic energy can exist in our reality it must be a force – thoughts are things'), to mould the future, to the Gaia hypothesis that the planet Earth is a living organism.

SAGE (SEMI AUTOMATIC GROUND ENVIRONMENT)

A ground/air radar interfaced with a computer whereby decisions can be made to launch Air Defence Systems.

LOCKHEED MARTIN M12/D21-A

Photograph 11 shows the Lockheed Martin M12 carrying the D21-A high speed reconaissance uninhabited vehicle which had a design speed of around 3,220kmph (2,000mph) and an operational ceiling of around 24,400m (80,000ft). The D21 utilized stealthy coatings on its surfaces and in that regard was the precursor to the F117 fighter and the B2 bomber. The M12 was the CIA variant of the SR71 'Blackbird' built for the USAF. The 'piggyback' project was discontinued following a fatal accident but continued with the D21 underslung and launched from a modified B52 bomber until early 1971. However, no doubt due to technical command difficulties, or hostile action, the D21 was lost in the pursuance of a mission. It was never flown again. The D21 was a remarkable achievement and was probably hindered greatly by being in advance of its time.

CANADAIR CL 227 'SENTINEL' VTOL UAV

This device, which has been developed over 20 years has a contra rotating turboshaft engine and is a front runner in the lucrative maritime theatre of surveillance requirements. Its strange shape, flight characteristics and use of non-metallic composites ensures high operational survivability. (See photograph 12.)

TIER DESIGNATIONS

The 'Tier' designations are ARPA procurement project names (Advanced Research Projects Agency – originators of the stealth fighter aircraft) and are facilitated by US Federal Law which gave DARPA (Defense Advanced Research Projects Agency – a think-tank dedicated to pushing the envelope of advanced technologies without the burden of bureaucracy or inter-service rivalries) special powers under Section 845 – other agreements granted authority to ARPA by Congress for prototype development outside the normal channels of Pentagon procurement procedures. In that respect the 'Tier' designation represents the pecking order of UAV projects. Current Tier projects include Tier I 'Predator', the Tier II Plus (constructed by TeleDyne Ryan in San Diego, California) which can fly at altitudes of 19,800m (65,000ft) for 24 hours or more, and the Tier III Minus 'Darkstar' which represents current state of the art surveillance technology, and is allegedly more stealthy than Tier II Plus. However, it is clear that no attempt has been made to minimize radar cross-sections by faceting surfaces or the jet intake. Like most radical modern aeronautical designs, the vehicle relies heavily on its on-board computers to sustain normal flight. (See photograph 13.)

GENERAL ATOMICS AERONAUTICAL SYSTEMS 'PREDATOR'

This UAV is equipped with technical wizardry which gives it an optical range of some 2,090km (1,300 miles) at a cruising height of 7,600m (25,000ft). Its first flight was in 1994 and it was successfully used in the Bosnian crisis. A big advantage of the package is the methodology employed to disseminate 'real time' digital imagery via satellite comms links across friendly Continental divides bringing the system very near to commercial broadcasting techniques – except of course the information is restricted to military information gathering. (See photograph 14.)

LOCKHEED MARTIN F22 'RAPTOR'

The Lockheed Martin F22 Raptor ATF (Advanced Technical Fighter) is to be the USAF replacement for the McDonnell Douglas F15 Eagle and was rolled out in April 1997. The F22 Raptor represents the cutting edge of modern aviation technology. It is largely constructed of titanium alloys and composites including graphite-bismaleimide for skins and internals. Aluminium honeycomb structures are used in critical areas to ensure maximum airframe integrity in combat situations. New manufacturing techniques include abrasive water jet drilling of titanium vent screens for radar defeating purposes and resin transfer mouldings. Robot devices are also used to fill in small manufacturing gaps so as to reduce radar reflections at the painting stage. The F22 Raptor is built by Lockheed Martin Aeronautical Systems and the Boeing Aircraft Corporations. It represents the current thinking on stealth and high survivability in a manned multi-role combat aircraft. The F22 Raptor is capable of around Mach 1.6 in level flight without afterburn. It has vectored thrust capability and a positive 'g' limit of around +8. It also carries extensive armaments and advanced mission avionics. (See photograph 15.)

UK's FOAS (FUTURE OFFENSIVE AIR SYSTEM) – MANNED AIRCRAFT STUDY

A manned aircraft study will define variants of 'Eurofighter' and the Tornado GR4 together with a new design interdictor aircraft optimised for FOAS roles. Phase I of the study was due for completion in January 1998 and selected concepts are to be studied in the second phase which is due to run until December 1999. A final decision must be made before the Tornado GR1/4 leaves service during 2015–18. The depicted artwork represents just one of the new concepts under consideration. Unmanned vehicles are also under consideration. (See photograph 16.)

BIBLIOGRAPHY AND FILMS

Barclay, David, *Aliens – The Final Answer* (Blandford Press, 1995)

Behrendt, Kenneth W., 'Introduction to Anti-mass Field Physics', *Aura* (August 1985, Vol 1, No 3)

Bord, Janet and Colin, *Life Beyond Planet Earth* (Grafton Books, 1991)

Brion, Marcel, *The World of Archaeology* (Elek Books, 1961)

Butler, Brenda, Street, Dot, and Randles, Jenny, *Sky Crash* (Grafton Books, 1984)

Collins, Tony, *Open Verdict – An Account of 25 Mysterious Deaths in the Defence Industry* (Sphere Books, 1990)

Disney, Michael, *The Hidden Universe* (J. M. Dent & Sons Ltd, 1984)

Flying Saucer Review Vol 31, No 3; Vol 38, No 3; Vol 36, No 2

Fortean Times, 'UFOs 1947–1987: The 40-Year Search for an Explanation' (Copyright BUFORA, 1987)

Fuller, John G., *The Interrupted Journey – Two Lost Hours Aboard a Flying Saucer* (Souvenir Press, 1966)

Good, Timothy, *Above Top Secret* (Sidgwick and Jackson Ltd, 1987)
    Alien Liaison – The Ultimate Secret (Random Century, 1991)
    Alien Update (Arrow Books, 1993)
    Alien Contact – Top Secret Files Revealed (William Morrow & Sons, 1993)

Hopkins, Budd, *Intruders – The Incredible Visitations at Copley Woods* (Sphere Books Ltd (UK), 1988)

Hopkins, Budd, Bloecher, Ted, and Clamar, Dr Aphrodite, *Missing Time: A Documented Study of UFO Abductions* (Marek, New York, 1981).

Jacobs, David M., *Secret Life – First-hand Accounts of UFO Abductions* (Simon & Schuster, 1992 (USA); Fourth Estate Ltd, 1993 (UK))

Jung, C. G., *Flying Saucers – A Modern Myth of Things Seen in the Sky* (Ark Paperbacks, 1977; first published 1959)

Kaku, Michio, *Hyperspace – A Scientific Odyssey Through Parallel Universes, Time Warps and the Tenth Dimension* (Oxford University Press, 1994)

Kennedy, Col W. V., Baker, Dr D., Friedman, Col R. S., and Miller, Lt Col D. *The Intelligence War* (Salamander Books Ltd, 1983)

Keyhoe, Donald E., *Flying Saucers from Outer Space* (Hutchinson & Co Ltd, 1954)

Kinder, Gary, *Light Years* (Penguin Books, 1987)

Kolosimo, Peter, *Not of This World* (Souvenir Press, 1970)

Leslie, Desmond, and Adamski, George, *Flying Saucers Have Landed* (Futura, 1953, 1977)

Mack, Prof John E., *Abductions – Human Encounters with Aliens* (Simon & Shuster, 1994)

Nichelson, Oliver, *Nikola Tesla's Long Range Weapon* (Vanguard Sciences, 1989)

Nichols, Preston B., and Moon, Peter, *Montauk Revisited – Adventures in Syncronicity* (Sky Books 1994)

Pond, Dale, *Energy Unlimited: Walter Baum Gartner* (Delta Spectrum Research Inc)

Randle, Kevin D., and Schmitt, Donald R., *UFO Crash at Roswell* (Avon Books, 1991)

Spencer, John, *Perspectives* (Macdonald & Co Ltd, 1989)

Stranges, Dr Frank E., PhD, *Stranger at the Pentagon* (Inner Light Publications; 1967, 1991)

Strieber, Whitley, *Majestic* (Macdonald and Co (Publishers) Ltd, 1990)

   Communion (Arrow Books Ltd, 1988)

   Transformation – The Breakthrough (Century Hutchinson Ltd, 1988)

Talbot, Michael,*The Holographic Universe* (Grafton Books, 1991)

Tomas, Andrew, *We Are Not The First* (Souvenir Press, 1971)

Vallée, Jacques, *Revelations – Alien Contact and Human Deception* (Souvenir Press, 1992)

Walters, Ed and Francis, *UFOs: The Gulf Breeze Sightings* (Bantam Press, 1990)

Wilson, Don, *Secrets of our Spacehip Moon* (Sphere Books Ltd, 1980)

FILMS AND TAPES

Hanger 18 Taft International Pictures Inc, 1980 (distributed by Worldvision Enterprises Inc, 1985)
Roswell – The Film Republic Pictures 1994 (distributed by Viacom Pictures)
The Lazar Tape Tri-Dot Productions Ltd, 1324 S. Eastern, Las Vegas, Nevada 89104, USA

INDEX